Antioxidants in Cocoa

Antioxidants in Cocoa

Editors

Dorota Żyżelewicz
Joanna Oracz

MDPI • Basel • Beijing • Wuhan • Barcelona • Belgrade • Manchester • Tokyo • Cluj • Tianjin

Editors
Dorota Żyżelewicz
Lodz University of Technology
Poland

Joanna Oracz
Lodz University of Technology
Poland

Editorial Office
MDPI
St. Alban-Anlage 66
4052 Basel, Switzerland

This is a reprint of articles from the Special Issue published online in the open access journal *Antioxidants* (ISSN 2076-3921) (available at: https://www.mdpi.com/journal/antioxidants/special_issues/Antioxidants_Cocoa).

For citation purposes, cite each article independently as indicated on the article page online and as indicated below:

LastName, A.A.; LastName, B.B.; LastName, C.C. Article Title. *Journal Name* **Year**, *Volume Number*, Page Range.

ISBN 978-3-0365-0232-8 (Hbk)
ISBN 978-3-0365-0233-5 (PDF)

Contents

About the Editors

Dorota Żyżelewicz is the Professor of Lodz University of Technology (TUL). She is the Deputy Director for Education at the Institute of Food Technology and Analysis of TUL and the Team Leader of Starch and Confectionery Technology. She is also the chair and member of many faculty and university commissions, the member of the Committee of Food Sciences and Nutrition of the Polish Academy of Sciences and the expert in the Supreme Technical Organization. She has published scientific articles in peer-reviewed journals, chapters in books, patents and has issued opinions on technology innovation and the implementation of an innovative technology or product. She was the leader, principal investigator or investigator of many scientific projects (grants). Her research interests focus on technology and analysis of food, bioactive components of plant food (e.g. polyphenols, tocopherols, phytosterols, methylxantines, Maillard reaction products, antioxidant activity, isolation and characterization of bioactive compounds, nutraceuticals, stability of food, storage, confectionery technology, cocoa bean, roasting, chocolate, functional food).

Joanna Oracz is Assistant Professor in Institute of Food Technology and Analysis at the Faculty of Biotechnology and Food Sciences of Lodz University of Technology. Her research interests focus on the development of new techniques for the isolation, purification, identification and structure characterization of phytochemicals with anti-oxidant, anti-inflammatory and anti-cancer activity, and the impact of processing on the bioactive components especially polyphenols and Maillard reaction products, and functional properties of selected plant materials. She is author or co-author of more than 39 publications in international, peer-reviewed journals in the field of food technology and nutrition.

 *antioxidants*

Editorial

Antioxidants in Cocoa

Joanna Oracz * and Dorota Żyżelewicz

Institute of Food Technology and Analysis, Faculty of Biotechnology and Food Sciences,
Lodz University of Technology, 4/10 Stefanowskiego Street, 90-924 Lodz, Poland; dorota.zyzelewicz@p.lodz.pl
* Correspondence: joanna.oracz@p.lodz.pl; Tel.: +48-42-631-3462

Received: 1 December 2020; Accepted: 2 December 2020; Published: 4 December 2020

Cocoa beans are the seeds of the tropical tree *Theobroma cacao* L. Because of the high concentration of bioactive compounds, including antioxidants (polyphenols, tocopherols), they are valued not only in the food industry but also in the pharmaceutical and cosmetic ones [1]. In recent years, interest in these cocoa components has greatly increased because of their potentially beneficial effects on human health. Cocoa antioxidants can inhibit or delay cellular damage either by quenching free radicals or through chelation of transition metal ions, which reduces their capability to form reactive oxygen species. They also exhibit a wide range of physiological properties resulting in protection against diseases, including coronary heart diseases, cancer or neurodegenerative disorders [2–5].

This Special Issue consists of 10 articles related to the effects of genotype and processing conditions on the phenolic compounds profile and antioxidant activity of cocoa derived products, isolation and characterization of antioxidant compounds such as polyphenols and melanoidins from cocoa beans, and assessment of the antioxidant, anti-oxidative stress and anti-inflammatory effects of cocoa beans and cocoa-derived products.

Several studies have well established that processing of cocoa bean including fermentation, drying, alkalization and roasting caused considerable changes in the chemical composition of the final product such as cocoa powder or chocolate [1,6–8]. During processing of the cocoa beans, the naturally occurring antioxidants (polyphenols) undergo significant changes in their structure that may impact their bioactivities. Fermentation and drying of the cocoa beans lead to oxidative degradation of polyphenols as a result of contact with the oxidative enzymes, polyphenol oxidase (PPO) and peroxidase. Monomeric flavan-3-ols are enzymatically oxidized to semi-quinones and quinones. Furthermore, these oxidation products are polymerized to condensed high molecular weight insoluble tannins. The native polyphenols may also react with proteins, free amino acids, and mono- or polysaccharides during roasting to form new compounds with antioxidant activities. The major chemical pathways which occur during roasting of cocoa beans and lead to the formation of new molecules are Maillard reactions [1,6,9]. Toro-Uribe et al. [7] focused their research on the understanding the mechanism of inhibition of PPO in cocoa beans, to find the optimal conditions, like concentration of inhibitor, temperature, and time, which enhance inhibition of PPO in cocoa beans without decreasing cocoa polyphenol concentrations. Their results showed that the optimum conditions to obtain the lowest PPO activity and highest total polyphenol content were achieved with 70 mM inhibitory solution (ascorbic acid/L-cysteine) at 96 °C for 6.4 min. Moreover, the described results evidence that heat treatment is a fast and robust method to reduce the activity of PPO enzyme in cocoa beans. As a result, the authors stated that this procedure also increases the extraction yield of polyphenols, and thus can be used to obtain enriched polyphenol extract with low PPO activity. Racine et al. [8] studied the effect of fermentation and roasting on the composition and α-glucosidase inhibitory activity of cocoa beans and powder, and identify the compositional factors and processing conditions that optimize α-glucosidase inhibitory activity of cocoa. They confirmed that processed cocoa powders are promising inhibitors of α-glucosidase, despite a significant reduction in total polyphenol and flavanol concentrations during fermentation and roasting. Due to this, the authors conclude that cocoa processing might generate compounds which enhance cocoa bioactivity, such as Maillard reaction products (MRPs), most notably melanoidins. Fernández-Romero et al. [9] investigated the degradation kinetics parameters of polyphenol and monomeric flavan-3-ols (catechin and epicatechin) during the roasting process of *Criollo* cocoa. The results indicate that degradation kinetics of the total phenolic content and epicatechin showed first-order reactions as the temperature increases, while the catechin showed patterns of formation and degradation. The authors also conclude that roasting at moderate temperatures is necessary to obtain minimal degradation of cocoa polyphenols and consequently antioxidant activities. Urbańska and Kowalska [10] focused their research on the comparison of the polyphenols content and antioxidant activity of chocolates produced from roasted and unroasted cocoa beans from different origins (Ghana, Venezuela, the Dominican Republic, Colombia and Ecuador). The findings demonstrated that the content of polyphenols vary greatly and depends on many factors, both those resulting from the genotype and the geographical and environmental conditions during growth, as well as the technological processes and parameters used. The obtained results indicate that both the beans (roasted and unroasted) and chocolates produced from them exhibited strong free radical-scavenging activity in vitro. In another study, Urbańska et al. [11] examined the effect of conching on the antioxidant potential of chocolate milk masses, taking into account different protein contents in milk obtained by spray or cylindrical drying. The results demonstrate the association between the protein content of milk powder and cocoa mass and the antioxidant potential of chocolate milk masses after conching. The results of these studies show that it is possible to maintain or increase the biological activity of cocoa beans and their derived products (cocoa powder and chocolate) by choosing appropriate processing conditions and cocoa genotype and origin.

Many recent studies revealed that cocoa beans and cocoa-derived products could be considered as an attractive source material for manufacturing of functional foods and nutraceuticals due to their very high content of bioactive compounds, mainly polyphenols, including flavonoids (proanthocyaninidins, monomeric flavan-3-ols, and anthocyanins) and phenolic acids, as well as melanoidins [1–5,7–9]. Toro-Uribe et al. [12] focused their research on developing a food-grade and suitable procedure for the recovery of polyphenols from cocoa beans avoiding the degreasing process. The results showed that concentration of ethanol, pH, temperature, irradiation time, and solid-to-solvent ratio affected significantly the yield of methylxanthines, catechins, and procyanidins with a degree of polymerization up to seven, as well as high antioxidant activity determined by oxygen radical absorbance capacity (ORAC). The optimal extraction conditions were 50% (*v/v*) ethanol, pH 6, 70 °C, and 45 min at the solid-to-solvent ratio of 1:120 *w/v*. Thus, they found that ultrasound-assisted solid–liquid extraction is a suitable method for the recovery of cocoa polyphenols and the obtained cocoa extract can be used as a valuable ingredient for functional food, nutraceuticals and cosmetics. Another study that investigated the total phenolic content, antioxidant properties, and structure–activity relationships of high-molecular weight (HMW) melanoidin fractions isolated by dialysis (>12.4 kDa) from raw and roasted under different temperature and relative air humidity conditions, cocoa beans of *Criollo*, *Forastero*, and *Trinitario* beans cultivated in different origins was conducted by Oracz and Żyżelewicz [13]. The results showed that it is possible to enhance the in vitro antioxidant properties of HMW cocoa melanoidins by choosing the appropriate roasting conditions and cocoa type. Moreover, structural analysis confirmed the presence of different bioactive compounds with various mechanisms of action in HMW cocoa melanoidin fractions. These results revealed that the HMW melanoidins fraction from roasted beans of different cocoa types could be considered as a valuable functional ingredient due to its high antioxidant properties in vitro (e.g., reducing power, antioxidant capacity, chelating activity) and total phenolic contents.

Felice et al. [14] paid attention to the antioxidant effect of cocoa husk extract and cherry extract against reactive oxygen species (ROS)-induced oxidative stress in Human Umbilical Vein Endothelial Cells (HUVECs). The results indicate that polyphenols in both extracts are effective in inhibiting ROS. Interestingly, it was also demonstrated that cocoa husk extract possesses an antioxidant effect even at low concentrations. In particular, the authors demonstrated that cocoa husk extract exhibited greater performance on HUVECs and had higher permeability across the rat intestine compared to cherry extract. The results clearly indicate that cocoa husk extract can be utilized as a valuable and cheap source of antioxidants with ROS scavenging ability that may have great relevance in the prevention of cardiovascular diseases (CVD).

The animal model studies further point to the therapeutic potential of both cocoa extract and cocoa polyphenols in metabolic and cardiovascular alterations. A study reported by Kluknavsky et al. [15] describes the genomic effects of the epicatechin during the stimulation of nitric oxide (NO) release and antioxidant defense in the aorta and left heart ventricle (LHV) investigated by using young male borderline hypertensive rats during two weeks of treatment (100 mg/kg/day p.o.) and two weeks post treatment. The obtained results indicate that a two-week oral administration of epicatechin decreased significantly blood pressure of young male borderline hypertensive rats, and these effects persisted for two weeks after the cessation of the treatment. The authors concluded that the mechanism of the anti-hypertensive effects of epicatechin is considered to be due to the reduced relative content of the iron-containing compounds in the blood, reduced $O_2^{\bullet-}$ production, and increased nitric oxide synthase (NOS) activity in the aorta and LHV. Finally, Ahmed et al. [16] studied the influence of cocoa flavanols on myocardial injury following acute coronary ischemia-reperfusion. The results demonstrated that 15-day oral administration of cocoa extract containing 250 mg/g flavan-3-ols (procyanidin) to Sprague–Dawley rats protects against myocardial ischemia-reperfusion (I/R) injury and significantly attenuates nitro-oxidative stress, inflammation, and mitigates myocardial apoptosis. It is well established that oxidative stress is considered to reflect an intracellular redox imbalance between the pro- and anti-oxidants, and plays a fundamental role in the pathogenesis of various diseases [15,16].

Thus, the findings demonstrated that a diet high in cocoa antioxidants could provide beneficial effects against risk factors of CVD, cancer, and neurodegenerative disorders. In vitro and in vivo studies reported the importance of cocoa antioxidants for the prevention of oxidative stress and inflammation. However, further clinical trials in humans are needed to confirm the health benefits of cocoa antioxidants.

Funding: This research received no external funding.

Conflicts of Interest: The authors declare no conflict of interest.

References

1. Kongor, J.E.; Hinneh, M.; de Walle, D.V.; Afoakwa, E.O.; Boeckx, P.; Dewettinck, K. Factors influencing quality variation in cocoa (*Theobroma cacao*) bean flavour profile—A review. *Food Res. Int.* **2016**, *82*, 44–52. [CrossRef]
2. Andújar, I.; Recio, M.C.; Giner, R.M.; Ríos, J.L. Cocoa polyphenols and their potential benefits for human health. *Oxid. Med. Cell. Longev.* **2012**, *2012*. [CrossRef] [PubMed]
3. Jaramillo Flores, M.E. Cocoa flavanols: Natural agents with attenuating effects on metabolic syndrome risk factors. *Nutrients* **2019**, *11*, 751. [CrossRef] [PubMed]
4. Jalil, A.M.M.; Ismail, A. Polyphenols in cocoa and cocoa products: Is there a link between antioxidant properties and health? *Molecules* **2008**, *13*, 2190–2219. [CrossRef] [PubMed]
5. Arranz, S.; Valderas-Martinez, P.; Chiva-Blanch, G.; Casas, R.; Urpi-Sarda, M.; Lamuela-Raventos, R.M.; Estruch, R. Cardioprotective effects of cocoa: Clinical evidence from randomized clinical intervention trials in humans. *Mol. Nutr. Food Res.* **2013**, *57*, 936–947. [CrossRef] [PubMed]
6. Barišić, V.; Kopjar, M.; Jozinović, A.; Flanjak, I.; Ačkar, Đ.; Miličević, B.; Šubarić, D.; Jokić, S.; Babić, J. The chemistry behind chocolate production. *Molecules* **2019**, *24*, 3163. [CrossRef] [PubMed]
7. Toro-Uribe, S.; Godoy-Chivatá, J.; Villamizar-Jaimes, A.; Perea-Flores, M.; López-Giraldo, L. Insight of polyphenol oxidase enzyme inhibition and total polyphenol recovery from cocoa beans. *Antioxidants* **2020**, *9*, 458. [CrossRef] [PubMed]

8. Racine, K.; Wiersema, B.; Griffin, L.; Essenmacher, L.; Lee, A.; Hopfer, H.; Lambert, J.; Stewart, A.; Neilson, A. Flavanol polymerization is a superior predictor of α-glucosidase inhibitory activity compared to flavanol or total polyphenol concentrations in cocoas prepared by variations in controlled fermentation and roasting of the same raw cocoa beans. *Antioxidants* **2019**, *8*, 635. [CrossRef] [PubMed]

9. Fernández-Romero, E.; Chavez-Quintana, S.; Siche, R.; Castro-Alayo, E.; Cardenas-Toro, F. The kinetics of total phenolic content and monomeric flavan-3-ols during the roasting process of Criollo cocoa. *Antioxidants* **2020**, *9*, 146. [CrossRef] [PubMed]

10. Urbańska, B.; Kowalska, J. Comparison of the total polyphenol content and antioxidant activity of chocolate obtained from roasted and unroasted cocoa beans from different regions of the world. *Antioxidants* **2019**, *8*, 283. [CrossRef] [PubMed]

11. Urbańska, B.; Szafrański, T.; Kowalska, H.; Kowalska, J. Study of polyphenol content and antioxidant properties of various mix of chocolate milk masses with different protein content. *Antioxidants* **2020**, *9*, 299. [CrossRef] [PubMed]

12. Toro-Uribe, S.; Ibañez, E.; Decker, E.; Villamizar-Jaimes, A.; López-Giraldo, L. Food-safe process for high recovery of flavonoids from cocoa beans: Antioxidant and HPLC-DAD-ESI-MS/MS analysis. *Antioxidants* **2020**, *9*, 364. [CrossRef] [PubMed]

13. Oracz, J.; Zyzelewicz, D. In vitro antioxidant activity and FTIR characterization of high-molecular weight melanoidin fractions from different types of cocoa beans. *Antioxidants* **2019**, *8*, 560. [CrossRef] [PubMed]

14. Felice, F.; Fabiano, A.; De Leo, M.; Piras, A.; Beconcini, D.; Cesare, M.; Braca, A.; Zambito, Y.; Di Stefano, R. Antioxidant effect of cocoa by-product and cherry polyphenol extracts: A comparative study. *Antioxidants* **2020**, *9*, 132. [CrossRef] [PubMed]

15. Kluknavsky, M.; Balis, P.; Skratek, M.; Manka, J.; Bernatova, I. (–)-Epicatechin reduces the blood pressure of young borderline hypertensive rats during the post-treatment period. *Antioxidants* **2020**, *9*, 96. [CrossRef] [PubMed]

16. Ahmed, S.; Ahmed, N.; Rungatscher, A.; Linardi, D.; Kulsoom, B.; Innamorati, G.; Meo, S.; Gebrie, M.; Mani, R.; Merigo, F.; et al. Cocoa flavonoids reduce inflammation and oxidative stress in a myocardial ischemia-reperfusion experimental model. *Antioxidants* **2020**, *9*, 167. [CrossRef] [PubMed]

Publisher's Note: MDPI stays neutral with regard to jurisdictional claims in published maps and institutional affiliations.

Article

Insight of Polyphenol Oxidase Enzyme Inhibition and Total Polyphenol Recovery from Cocoa Beans

Said Toro-Uribe [1], Jhair Godoy-Chivatá [1], Arley René Villamizar-Jaimes [2], María de Jesús Perea-Flores [3] and Luis J. López-Giraldo [1,*]

[1] School of Chemical Engineering, Food Science & Technology Research Center (CICTA), Universidad Industrial de Santander, Carrera 27, Calle 9, 68002 Bucaramanga, Colombia; saidtorouribe@gmail.com (S.T.-U.); jhaigo@hotmail.com (J.G.-C.)

[2] Food Science & Technology Research Center (CICTA), Universidad Industrial de Santander, Carrera 27, Calle 9, 68002 Bucaramanga, Colombia; arleyvil@uis.edu.co

[3] Centro de Nanociencias y Micro y Nanotecnologías, Instituto Politécnico Nacional, Luis Enrique Erro s/n, Unidad, Profesional Adolfo López Mateos, Col. Zacatenco, C.P. 07738 Ciudad de México, Mexico; mpereaf@ipn.mx

[*] Correspondence: ljlopez@uis.edu.co; Tel.: +57-300-377-8801

Received: 1 March 2020; Accepted: 28 April 2020; Published: 27 May 2020

Abstract: A full factorial design (ascorbic acid/L-cysteine inhibitors, temperature, and time as factors) study was conducted to enhance inhibition of polyphenol oxidase (PPO) activity without decreasing cocoa polyphenol concentrations. The data obtained were modelled through a new equation, represented by Γ, which correlates both high polyphenol content with reduced specific PPO activity. At optimized values (70 mM inhibitory solution at 96 °C for 6.4 min, Γ = 11.6), 93.3% PPO inhibition and total polyphenol of 94.9 mg GAE/g were obtained. In addition, microscopy images confirmed the cell morphological changes measured as the fractal dimension and explained the possible cell lysis and denaturation as a result of heat treatment and chemical inhibitors. Results also showed that PPO enzyme was most suitable (higher v_{max}/K_m ratio) for catechol, with a reduction in its affinity of 13.7-fold after the inhibition heat treatment. Overall, this work proposed a suitable and food-safe procedure for obtaining enriched polyphenol extract with low enzyme activity.

Keywords: polyphenol oxidase; cocoa polyphenols; heat treatment; enzyme inactivation

1. Introduction

Precursors of chocolate flavor are usually obtained through enzymatic and non-enzymatic reactions, in which polyphenol oxidase, invertase, and protease are the most important enzymes [1]. Polyphenol oxidase (PPO) is a major copper enzyme [2], also known as catechol oxidase, tyrosinase, and so forth [3], and is the most important deteriorative enzyme that accelerates oxidation and degradation of polyphenols and their derivatives [4]. PPO is located in the chloroplasts [5,6] and its activation takes places during cell-damaging treatment (e.g., slicing, cutting or pulping) where oxygen is available and the local pH is not too acidic [7,8], thus causing the formation of brown pigments [2,9]. In fact, the oxygen catalyzes the enzyme reaction where the monophenols forming *o*-diphenols (monophenolase activity) and then the oxidized substrate react, producing *o*-quinones (diphenolase activity) [10,11].

The rate of enzymatic browning on food is governed by PPO action, which depends on concentration, pH, temperature, amount of phenolic compounds, and oxygen availability [3,12], thus having a different level of influence on the development of flavor, color and softening, which in turn is reflected in the loss of nutritional and quality value [13]. For instance, Misnawi et al. [7] reported that PPO of dried unfermented beans increases the polyphenol oxidation rate of (−)-epicatechin, total polyphenols, and total anthocyanidins. PPO is also susceptible during the fermentation stage; therefore, total and specific activities remaining in unfermented beans are reduced up to 1% and 9% of the original, respectively [7]. Despite the strong inactivation of PPO during fermentation, it can be regenerated during the drying process via –pH increase and uptake of O_2^-, and the remaining PPO activity is sufficient to catalyze oxidation of phenolic compounds [12,14]. Thereby, phenolic compound content in cocoa is affected by several factors: not only by the genetic origin, geographical and environmental conditions, but also by enzyme attack and processing conditions for chocolate production.

Polyphenolic compounds have been widely studied, since they possess an array of nutraceutical properties for human health related to cardiovascular effects [15], antioxidant activity [16], anti-inflammatory response [17], antibactericidal effect and biological applications [18,19]. As a result of all these functional bioactivities, enriched polyphenol extracts have gained greater attention. For instance, new products such as exGrape®SEED, Vitaflavan from grape seeds, enriched capsules with high amounts of cocoa procyanidins (PCs), enriched dark chocolate bars (e.g., CocoaVia from Mars, and FlavaBars® by Flavanaturals), and Fulyzaq™ for antiretroviral-induced diarrhea (PCs consisting of DP 3–30 from *Croton lechleri*; FDA approval) are used as a food supplement. In this sense, research focusing on inactivation of enzymes without affecting the total polyphenol content deserves further investigation.

Secondary metabolites of cocoa are purine alkaloids (e.g., caffeine, theobromine, and theophylline) and flavan-3-ols, which comprise between 0.05–1.7 wt %, and 12–18 wt % [20,21], respectively. The (−)-epicatechin constitutes the major monomeric flavanol, which also forms oligomeric and polymeric procyanidins of (epi)catechin units up to tridecamers [22]. The flavan-3-ols are also characterized as including OH groups in *ortho* position, which make an excellent substrate for PPO.

Regarding inactivation of PPO and its relationship with polyphenols, it has been studied in apricots, apples, grapes, tea leaves, potatoes, lettuce, coffee, black raisins, anthocyanidins from strawberries, catechins, quercetin, shrimps, and others [10,11]. Several inactivation strategies, for instance, using reacting enzymes [23], reducing agents (e.g., removal of oxygen using chemical agents) [2], changes on pH [24], and increasing temperature [25] have been tested. Regarding reducing agents, sulfites have been widely employed, but, currently, their use has been restricted because of their adverse effects on human health [26]. Other anti-browning agents can be used, but only a limited number are considered acceptable in terms of safety and cost to control the enzymatic browning in foods or food products [27].

L-cysteine and ascorbic acid are the most widely used inhibitor agents [24]. In fact, Pizzocaro found that ascorbic acid at 0.01–56.8 mM acted as an antioxidant reducing *o*-quinone back to the original phenol compound, while L-cysteine at 40–100 mM [2] made it possible to form complexes with *o*-quinones, and at 0.20–2.0 mM [28] it exhibited an inhibition of PPO higher than 98%. In addition, commission regulation EU No. 1129/2011 approved the use of ascorbic acid (food additive E-300) and L-cysteine (food additive E-920) in foods.

Regarding the inhibition of PPO in cocoa beans, heat inactivation at temperatures ranging from 60–98 °C, for a period ranging from 3–45 min, have been previously assayed [25,29]. For instance, Pons-Andreu et al. [29] proposed an enzymatic treatment for cocoa nibs using a blanching process. However, heating is not a valid treatment to enhance long-term inhibition of PPO activity, since the enzyme is highly thermostable [30]. In addition, several works evaluate the change of enzyme activity by qualitative color measurement (melanosis index scale) instead of measuring the specific enzymatic activity [29,31]. Furthermore, many questions remain unsolved concerning the inactivation process. None of the abovementioned studies investigated the optimal temperature, time of heat treatment, type or concentration of chemical inhibitors to enhance lower enzyme activity in cocoa beans. To our knowledge, the relationship between the total polyphenol content during the PPO denaturation and how this affects antioxidant capacity and the bioactive properties of cocoa beans has not been studied yet.

Therefore, the aims of this work were to: (a) determine the conditions to inhibit the PPO in cocoa beans while maintaining a high level of polyphenols (to do so, concentration of inhibitor (ascorbic acid and L-cysteine), temperature, and time were evaluated); (b) develop an equation showing the relationship between PPO inactivation and polyphenol content; and (c) investigate the enzyme's kinetic parameters and their affinity to PPO using catechol, (+)-catechin, and (−)-epicatechin as substrates.

2. Materials and Methods

2.1. Reagents

All the chemicals used were analytical reagent grade and were used without further purification. Folin–Ciocalteu reagent, gallic acid, sodium carbonate, catechol, bovine serum albumin, ascorbic acid, L-cysteine, sodium phosphate dibasic, citric acid, poly(vinylpyrrolidone) (PVP), and Coomassie brilliant blue G-250 dye were obtained from Sigma Aldrich (St. Louis, MO, USA). (+)-Catechin (≥99%), (−)-epicatechin (≥99%), and procyanidin B2 were purchased from ChromaDex Inc. (Irvine, CA, USA). Acetonitrile (HPLC-grade), ethanol (analytical-grade), and formic acid were acquired from Merck (Merck, Germany). Deionized water (18 MΩcm^{-1}) from an Aqua Solution system (Aqua Solution, Inc., Jasper, GA, USA) was used for the preparation of all solutions.

2.2. Recovery of Cocoa Polyphenol Extract

Fresh cocoa pods (Trinitary, clone ICS 39) were collected at Villa Santa Monica (San Vicente de Chucurí, Santander, Colombia) and immediately protected from light and transported on ice to CICTA Lab for processing. Cocoa pods are mainly composed of cocoa husk, cocoa beans, and mucilage. Thus, the cocoa beans were removed manually using a knife and the beans surrounded by mucilage were immediately removed using a mucilage remover (Penagos Ltda, Bucaramanga, Colombia). After that, the beans were immediately inactivated (as described in Section 2.3) and used for further analysis.

2.3. Enzyme Inhibition

The inhibition of PPO enzyme in cocoa beans was enhanced by heat treatment. The samples were dipped in an aqueous inhibitory solution consisting of ascorbic acid and L-cysteine (1:1 *v/v* ratio) at same equimolar concentration. The assays were carried out as follows: 10 beans (ca. 25 g wet weight at ca. 4 °C) were immersed into 200 mL of inhibitory solution at different concentrations (0–50 mM), times (1–5 min) and temperatures (70–90 °C) in accordance with the combinations of surface design 2^3, which includes four replicates, a central point, and start points (Table 1). These levels of factors were chosen with the goal to maintain a high level of polyphenols; therefore, it is preferred to use a high temperature for a shorter time. Immediately after the inmersion of beans into the inhbitory solution, the samples were cooled in ice water for 30 min, and then rinsed (×3) again with deionized water (4 °C) to remove traces of ascorbic acid, and L-cysteine. Non-treated sample (fresh unfermented cocoa bean) was kept as control.

2.4. PPO Enzyme Extraction

The treated and non-treated beans were chopped into small pieces and homogenized. The enzyme extraction was done according to Babu et al. [32] with few modifications. Briefly, the chopped pieces were homogenized in cold extraction buffer (ratio 1:1.5 w/v, 0.01 M McIlvaine citric phosphate, pH 6.5, containing 1% PVP) during 2 min at max speed (Vortex Reax Control, Heidolph, Germany) and filtered by Whatman No. 1 filter paper (Whatman, Inc., Florham Park, NJ, USA). Homogenates were centrifuged (Heraeus, Megafuge 16R, Thermo Scientific, Waltham, MA, USA) at 10,000× g and 4 °C for 20 min. The resulting supernatant, called crude enzyme extract, was filtered again and used for further experiments.

Table 1. 2^3 full factorial surface design and experimental results for the inhibition of PPO enzyme and higher polyphenol content from cocoa beans.

Run	T (°C)	Time (min)	Inhibitor [mM]	Specific Activity [U_{PPO}/mg]	Total Polyphenol [mg GAE/g]	Γ *
1	90	1	50	8.97 ± 0.08	86.58 ± 2.53	3.74
2	70	5	0	18.48 ± 0.49	83.37 ± 1.45	1.68
3	63	3	25	8.40 ± 0.45	80.52 ± 3.21	3.45
4	90	1	0	20.42 ± 0.05	70.48 ± 0.90	1.05
5	90	5	0	12.30 ± 0.52	84.74 ± 4.65	2.61
6	80	3	67	6.86 ± 0.99	88.20 ± 3.43	5.07
7	80	3	25	15.05 ± 0.49	69.24 ± 1.22	1.36
8	80	3	25	18.42 ± 0.66	75.84 ± 1.34	1.38
9	80	1	25	20.08 ± 0.52	63.62 ± 4.50	0.81
10	70	1	0	21.11 ± 0.06	53.65 ± 2.32	0.41
11	70	1	50	21.15 ± 0.19	63.11 ± 1.01	0.75
12	80	6	25	5.60 ± 0.47	87.68 ± 3.21	6.14
13	70	5	50	5.17 ± 0.82	85.08 ± 2.21	6.27
14	80	3	25	14.03 ± 0.70	66.52 ±2.56	1.31
15	90	5	0	12.04 ± 0.41	85.58 ± 6.65	2.72
16	97	3	25	5.09 ± 0.36	86.32 ± 3.27	6.55
17	90	5	50	6.22 ± 0.90	85.21 ± 4.05	5.23
18	97	3	25	5.00 ± 0.31	86.73 ± 2.41	6.74
19	80	3	25	14.94 ± 0.07	68.57 ± 4.10	1.34
20	70	5	50	5.50 ± 1.14	87.67 ± 1.43	6.25
21	80	3	0	20.01 ± 0.23	71.01 ± 1.32	1.09
22	80	1	25	22.16 ± 1.14	65.75 ± 0.53	0.81
23	63	3	25	8.69 ± 0.60	79.99 ± 4.32	3.29
24	90	5	50	6.30 ± 0.41	85.15 ± 4.21	5.15
25	80	6	25	5.43 ± 0.16	87.50 ± 5.01	6.30
26	70	1	0	18.09 ± 0.04	53.24 ± 2.45	0.46
27	80	3	0	23.99 ± 0.79	74.21 ± 1.23	1.01
28	90	1	50	8.62 ± 0.86	85.60 ± 3.02	3.81
29	90	1	0	21.36 ± 0.43	74.44 ± 2.03	1.14
30	70	1	50	24.22 ± 0.26	65.94 ± 2.01	0.74
31	80	3	67	5.93 ± 0.33	82.54 ± 2.98	5.14
32	70	5	0	17.54 ± 0.78	80.84 ± 3.89	1.67

* Γ calculated according to Equation (5).

2.5. PPO Enzyme Activity Measurement

The enzyme activity (U_{PPO}) was determined spectrophotometrically according to Pizzocaro et al. [24]. The reaction mixture, containing 1.0 mL of catechol solution (0.175 M) and 2.0 mL of McIlvaine buffer pH 6.5, was added to 0.5 mL crude enzyme extract. The increase in absorbance at 420 nm (Genesys 20; Thermo Scientific-Fisher, Waltham, MA, USA) was recorded at intervals of 15 s up to 5 min at room temperature. The PPO activity was calculated by the slope of the linear portion of the curve absorbance vs. time. The enzyme activity corresponding to one unit of PPO activity was defined as the 0.001-unit change in absorbance per minute at 420 nm per mL of enzyme assay solution mixture.

The protein content of specific activity was measured according to the method described by Bradford [33]. Bovine serum albumin (BSA) was used as a standard (0–0.125 mg/mL) ($r^2 = 0.999$). The specific activity was expressed as one unit of enzyme activity per one unit (mg^{-1}) of BSA protein (U_{PPO}/mg).

The percent of total inhibition was calculated as follows (Equation (1)):

$$Inhibition\ (\%) = \frac{Control - Test_i}{Control} * 100 \tag{1}$$

where i is the number of the treatment according to the design. *Control* and *test$_i$* were expressed as the amount of enzyme specific activity (U_{PPO}/mg).

2.6. Substrate Kinetic Constants of PPO

The evaluation of inhibition constant was assayed using catechol, (+)-catechin and (−)-epicatechin (main catechins in cocoa) as substrate (5–200 mM) at optimal temperature for PPO activity, that is, 35 °C, as previously reported in the literature [3,28].

The reaction was modeled using the Michaelis–Menten equation (Equation (2)). The K_m value and maximum velocity v_{max} were determined using a nonlinear regression by GraphPad Prism v. 6.0 (GraphPad Soft. Inc., La Jolla, CA, USA).

$$v = \frac{v_{max} * [s]}{K_m + [s]} \tag{2}$$

2.7. Recovery of Total Phenol Content

Recovery of phenolic compounds from non-treated cocoa beans (control sample) and beans remaining after the PPO inhibition treatment were determined as follows: beans were freeze-dried (Labconco Corp., Kansas City, MO, USA) for a final humidity <4% (according by AOAC method 931.04, 1990) [34], milled and homogenized (Grindomix GM 200, Retsch GmbH & Co., Haan, Germany). The extraction was carried out as follows: 1 g of sample was added to 60 mL of a mixture of 50% ethanol/water (w/w) at 50 °C, and stirred at 300 rpm for 30 min using a magnetic stirrer hotplate (IKA C-MAG HS7, Germany) and thermocouple (IKA ETS-D5, Germany). The resulting extract was centrifuged (5000× g, 4 °C, 20 min); then, the supernatant was filtered through 0.45 µm hydrophilic PTFE filter (Millipore, Milford, MA, USA), and immediately analyzed.

2.8. Determination of Total Polyphenol Content

The total polyphenol content of the sample was assayed using Folin–Ciocalteu reagent according to Singleton et al. [35] with modifications as follows: the reaction was initiated by the addition of 50 µL of the sample with 1.5 mL of 10-fold diluted Folin-Ciocalteu reagent. After 5 min, 1.5 mL of 7.5% (w/v) sodium carbonate was added and vortexed for 10 s. The reaction medium was stored in the dark for 1 h at 25 °C. Absorbance was measured at 765 nm (Genesys 20; Thermo Scientific-Fisher, Waltham, MA, USA) against a blank sample. A gallic acid calibration curve was prepared with 0.05–1.0 mg/mL ($r^2 = 0.999$). Results of total polyphenols amount were expressed as mg gallic acid equivalents by gram of dried cocoa beans (mg GAE/g).

2.9. Chromatographic Analysis by HPLC-DAD

The reverse phase conditions and stationary phase were optimized to detect both catechins and procyanidins in the cocoa extract. Briefly, LC was assayed on a Shimadzu (LC-2030 LT Series-i, USA) and the separation was achieved using a C18-phenyl column (4.6 × 50 mm, 2.5µm) (Xbridge, Waters, Milford, MA, USA) protected with a security guard from Phenomenex (AJ0-8788, Phenomenex, Torrance, CA, USA). The procedure consisted of water/formic acid (99.99/0.01 v/v) (solvent A), and acetonitrile/formic acid (99.99/0.01 v/v) (solvent B). The linear gradient was as follows: 0-8 min, 2% B; 8–37 min, 10% B; 37–40 min, 0% B and re-equilibrium for 10 min. The flow rate, column temperature, and diode array were set at 0.8 mL/min, 60 °C, and 280 nm respectively. Identification of both catechins and procyanidins were carried out by Ion Trap LC/MS (model 6320, Agilent Technologies, Waldbronn, Germany) equipped with an ESI source and ion trap mass analyzer, which was controlled by the 6300 series trap control software (Bruker Daltonik GmbH, V. 6.2). The mass spectrometer was operated in negative ESI mode with the following conditions: mass spectra recorded from 90–2200 m/z, nebulizer 40 psi, dry gas 12 L/min and dry temperature 350 °C. Target compounds consisted of $[M-H]^-$ ions at m/z 289, 577, 865, 1153, 1442, and 1730, which correspond to monomer, dimer, trimer, tetramer, pentamer, and hexamer procyanidins structures, respectively.

2.10. Scanning Electron Microscopy and Image Analysis

Scanning electron microscopy (SEM) was additionally used to evaluate the microstructure of (a) non-treated cocoa beans, and (b) cocoa beans with reduced PPO activity. Beans were cut into longitudinal and transversal sections with the objective to observe their microstructure. Sections were mounted on aluminum stubs with double-sided carbon adhesive tape and observed using the XL-30 Environmental Scanning Electron Microscope (Philips, Cambridge, MA, USA) at 25 kV accelerating voltage with a BSE (backscattered electron) detector. The images were acquired in grayscale and stored in TIFF format at 712 × 484 pixels.

Images of the samples were captured using electronic microscopy and stored as .TIF in a gray scale with brightness values between 0 and 255 for each pixel constituting the image. A generalization of the box-counting method to evaluate the fractal dimension of the images (FDt) was used. In this work, the shifting differential box-counting method was used (SDBC) [36] to evaluate the fractal dimension of texture of SEM images using the ImageJ 1.34 software. Four different images at the same magnification (1000×) were evaluated.

2.11. Statistical Analysis

All measurements were repeated at least three times. Statistical analysis was done using Statistica v. 7.1 (Stat-Soft Inc., Tulsa, OK, USA). The three-way analysis of variance (three-way ANOVA) and *p*-test were used to evaluate the influence of the factors and their interactions on the experimental design. One-way ANOVA and Tukey´s multiple range test at a 5% level of significance were also evaluated. The response surface methodology, consisting of a full factorial central composite rotatable design with four replicates at the central point was conducted according to a completely randomized model. A second-order polynomial equation was used to fit the experimental data, as follows (Equation (3)):

$$Y = \beta_0 + \sum_{i=1}^{k} \beta_i X_i + \sum_{i=1}^{k} \beta_{ii} X_i^2 + \sum_{\substack{i=1 \\ i<j}}^{k-1} \sum_{j=2}^{k} \beta_{ij} X_i X_j \tag{3}$$

where Y is the predicted factor, X is the extraction parameter, β_0 is the value of the fitted response to the design, and β_i, β_{ii}, and β_{ij} are the coefficients of linear, quadratic, and cross-product terms, respectively.

In this study, performance of full factorial central composite design by R-squared coefficient was measured. In addition, experimental runs were randomized to evaluate the concordance of experimental data and predicted values; therefore, the root-mean-square error (RMSE, Equation (4)) was used, as follows:

$$RMSE = \sqrt[2]{\frac{\sum_{i=1}^{n} (y_i - \hat{y}_i)^2}{n}} \tag{4}$$

where y_i and $\hat{y}_i$ is the measured value and predicted value by the model, respectively, and n is the number of the set data.

3. Results and Discussion

3.1. Preliminary Inhibition Assays

Prior to optimizing the inhibition of PPO enzyme from cocoa beans, the following parameters were evaluated: (a) the nature of the inhibitor, and (b) the size of the cocoa beans. Thus, the PPO activity as a function of different inhibitory agents was determined with a solution containing 1% (w/w) ascorbic acid, 1% (w/w) L-cysteine, and mixture of ascorbic acid/L-cysteine (1:1 ratio, 1% w/w,) using heat treatment at 90 °C for 5 min as previously reported by Menon et al. [25]. At concentration >1%, inhibitors behave as quinone reducers, similar to sulfides [26]. The results showed that highest denaturation of the enzyme was enhanced by a mixture of ascorbic acid/L-cysteine (79.3%), followed by ascorbic acid (72.8%) and L-cysteine (67.5%). In addition, two sizes of cocoa beans consisting of (S_1) chopped cocoa bean (cross section of 1×0.5 cm^2), and (S_2) whole cocoa beans, were also evaluated. Results showed that S_1 treatment inhibited the PPO 1.2-fold more than S_2. Interestingly, inhibition solution color after heat treatment (control system) was translucid-yellow, which was quite similar to S_2 treatment. However, a violet color in the waste solution of S_1 treatment was observed and could be a consequence to the greater surface contact, thus facilitating the release of polyphenols. Indeed, through HPLC-DAD-MS/MS, it was shown that loos of catechins and procyanidins (up to hexamers) on S_1 and S_2 treatments were 0.5 and 1.2, and 8 and 22 wt.%, respectively.

Hence, the maximization of the inactivation of PPO was carried out using whole cocoa beans, together with a combination of ascorbic acid/L-cysteine at same equimolar concentration.

3.2. Influence of Inhibition Parameters on PPO Activity

The extent of PPO inhibition as a function of treatment time, temperature and inhibitor concentration is summarized in Table 1. In addition, the recovery of total polyphenol content for each assay was evaluated (Table 1). As can be seen in Supplementary Figure S1, a non-linear relationship could be observed between values of enzyme inactivation and concentration of polyphenols ($r^2 = 0.60$). To better understand the relationship between the two response variables, several models, such as quadratic ($r^2 = 0.61$), exponential ($r^2 = 0.56$), and logarithmic ($r^2 = 0.59$) equations, were evaluated; however, none of them was able to describe the data adequately. Furthermore, a new equation (Equation (5)), which correlates high polyphenol content with reduced PPO activity—expressed in percentage—in an inverse relationship, was established as follows:

$$\Gamma_i = \frac{\text{Total Polyphenol}_i\ (\%)}{100 - \text{PPO Inhibition}_i\ (\%)} \tag{5}$$

where Γ—represented by the Greek uppercase letter—means high polyphenol content with low enzyme activity, as a function of % total polyphenol recovered and % PPO inhibition; i is the number of the treatment according to the design.

Figure 1 shows the experimental data adjusted according to our proposed model (Equation (5)). Plotting of Γ as a function of total polyphenol (%) or inactivation (%) produced good adjustments of r^2, equal to 0.91 and 0.92 respectively (Supplementary Figure S2A,B).

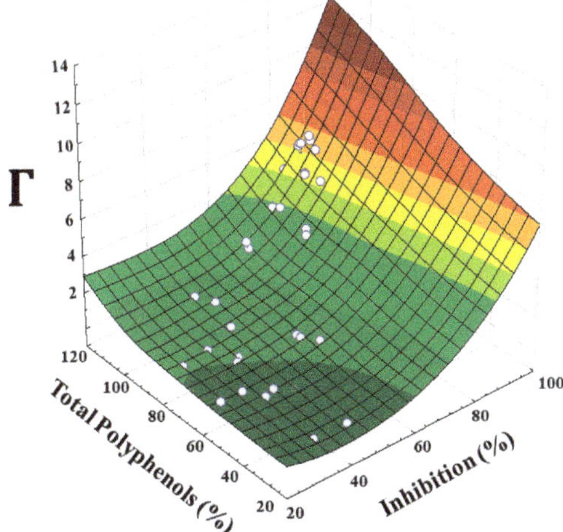

Figure 1. 3D scheme for the correlation Γ as a function of polyphenol oxidase (PPO) inhibition (%) and total polyphenol content (%) on cocoa beans. See Equation (5).

As can be seen, Γ had an exponential behavior; that is, with brief, low-heat treatment the rate of inactivation was lower, and both maximum enzyme inhibition and polyphenol content (lower thermal degradation of polyphenol compounds) were increased by increasing the temperature until a saturation value was reached. We hypothesized that heat treatment not only made it possible to break down the enzyme–substrate complex but also caused softening of the cell, and thereby increased the extraction yield of polyphenols.

3.3. Effects of Temperature, Time and Concentration of Inhibitor on PPO Activity

Analysis of variance shows that the selected quadratic model adequately represented the data obtained for Γ with a good coefficient of multiple determination of $r^2 = 0.891$ (Table 2) and lower residual values (MS residuals equal to 0.748). The model's ability to accurately predict the data based on randomly selected experiments ($n = 15$), by comparing how close predictions are to the actual outcomes, was assessed. Therefore, RMSE was 0.388, which reinforced the good performance of the model.

Table 2. ANOVA for polyphenol oxidase (PPO) inactivation through 2^3 surface design + central points + start points. $r^2 = 0.8083$; r^2adj = 0.8462.

Factor	SS	df	MS	p
T	8.3298	1	8.32985	0.003001
T^2	19.4339	1	19.43392	0.000042
t	49.5296	1	49.52962	0.000000
t^2	2.2011	1	2.20111	0.100515
Inh	35.4793	1	35.47934	0.000001
Inh^2	0.0185	1	0.01846	0.876667
T × t	3.5258	1	3.52580	0.041099
T × Inh	0.0242	1	0.02417	0.859066
t × Inh	4.2416	1	4.24158	0.026414
Error	16.4760	22	0.74891	
Total SS	150.9190	31		

SS is the sum of the squares, df is the degree of freedom, MS is the mean square, p is the probability value, T is temperature, t is time, and Inh is Inhibitor.

In general, ANOVA and the analysis of surface response (Table 2, Figure 2) confirmed the dependence of higher PPO denaturation as a function of the linear and quadratic effect of temperature and the linear effect of time of treatment, and concentration of inhibitor (Equation (6)). Besides, interactions between temperature with time and time with inhibitor concentration were also significant ($p < 0.05$). These can happen because (i) heat treatment affects the conformational change of the enzyme and protein–enzyme dissociation [37] and (ii) the dose-dependent inhibitory effect [2,28]. A similar trend was also observed by Oliveira and Orlanda [3], who found that the PPO enzyme was stable at temperature <67 °C, but greater denaturation of 90% could be enhanced at temperature >85 °C after 20 min [28].

$$Y = 54.649 - 1.496T + 0.010\,T^2 + 1.758\,t + 0.009Inh - 0.023\,T * t + 0.010\,t * Inh \tag{6}$$

At the maximum temperature (35 °C) for PPO activity, non-treated cocoa sample had a total polyphenol content of 42.1 mgGAE/g, a maximum PPO-specific activity of 32.0 ± 0.2 U_{PPO} mg^{-1}, and a protein amount of 68.5 ± 3.6 mg mL^{-1}, which was consistent with Lee et al. [37] and Misnawi et al. [7] with PPO-specific activity of 52 and 75 U_{PPO} mg^{-1}, and total protein of 70 and 73 mgmL^{-1}, respectively. Thereby, a significant effect of temperature ($p < 0.05$; Figure 2A,C) on the reduction of PPO activity, which was 1.95-, 2.23-, 2.66- and 6.35-fold lower at 70 °C, 80 °C, 90 °C, and 97 °C (compared to 35 °C), respectively, was observed. Regarding the length of treatment time (Figure 2B,C; Equation (6)), this showed the largest positive linear regression coefficient, suggesting that this factor is critical to enhancing the reduction of PPO activity. The effect of concentration of inhibitor (Figure 2A,B) was also significant ($p < 0.05$), indicating the role of ascorbic acid as an antioxidant reducing the initial *o*-quinone, and of L-cysteine as a reducing agent interfering with PPO activity before it can polymerize to melanin [3].

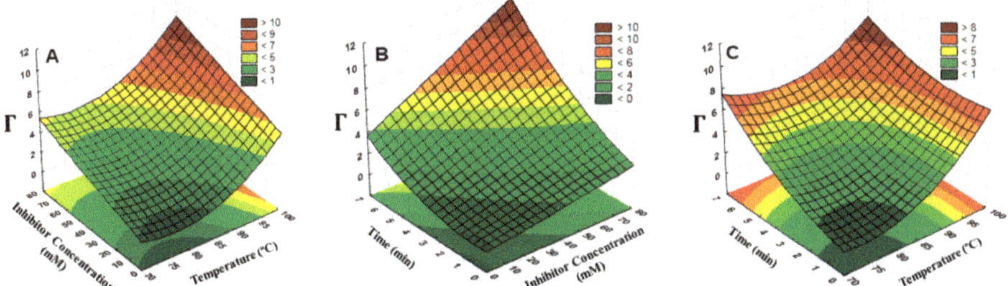

Figure 2. Surface response for the correlation of Γ with (**a**) temperature (T) and concentration of inhibitor (Inh), (**b**) time and Inh, (**c**) t and T. See Equation (3).

These findings reinforce the synergic effect of ascorbic acid and L-cysteine as an efficient solution to prevent enzymatic browning reactions. In fact, effective use of ascorbic acid and/or L-cysteine in combination with other compounds has been previously confirmed [28,38]. Based on our results, the optimum conditions to obtain the lowest enzymatic browning and highest total polyphenol content were achieved with 70 mM inhibitory solution at 96 °C for 6.4 min, for a predicted Γ = 11.8, which agreed with the experimental results obtained under these conditions, which provided a Γ = 11.6 ± 2.7, that is, 93.3 ± 2.1% PPO inhibition and total polyphenol of 94.9 ± 4.09 mg GAE/g (2.3-fold higher than non-treated samples). Under these conditions, a long-term study showed that inactivated cocoa beans (stored at 4 °C) maintain, over 2 years, their total phenol content (ca. 92 ± 3.2 mg GAE/g) and PPO activity (ca. 89 ± 3.8%) with no significant change over time ($p < 0.05$).

3.4. Kinetic Parameters of PPO Inhibition in Cocoa Beans

The differences in PPO activity observed varying the substrate (catechol, (+)-catechin, and (−)-epicatechin) were determined in the enzyme's kinetic parameters. As expected, all the substrates were oxidized, displaying Michaelis–Menten kinetics [8]. The kinetic parameters calculated by nonlinear regression are summarized in Table 3. Regarding the catalytic power (v_{max}/K_m ratio), taken as an evaluation criterion, the enzyme seemed to be most suitable for small *o*-diphenols such as catechol (4440.98 U mM^{-1} mL^{-1} min^{-1}), followed by (−)-epicatechin (727.38 U mM^{-1} mL^{-1} min^{-1}), and (+)-catechin (637.79 U mM^{-1} mL^{-1} min^{-1}). Indeed, higher affinity for catechol followed by catechin was also found by Doğru and Erat [2].

Table 3. Michaelis-Menten kinetic parameters on different substrates as action of cocoa bean PPO.

Substrate	K_m [mM]	v_{max} (U$_{PPO}$ mL^{-1} min^{-1})	v_{max}/K_m (U$_{PPO}$/mM mL min)
Catechol	0.61 ± 0.12 [a]	2709 ± 21.89 [a]	4440.98
Catechol *	8.36 ± 1.33 [b]	106.7 ± 3.70 [b]	12.76
(−)-Epicatechin	1.26 ± 0.37 [a]	916.5 ± 20.59 [c]	727.38
(+)-Catechin	1.45 ± 0.35 [a]	924.8 ± 18.98 [c]	637.79

* Substrate evaluated using PPO enzyme from inactivated cocoa beans at optimum conditions (70 mM inhibitory solution at 96 °C for 6.4 min). Means within a column sharing the same letter are not significantly different by Tukey ($p > 0.05$).

Affinities of PPO for catechins obtained in this study were quite similar to that reported by Wuyts et al. [39] and Ho [40] with K_m of 1.2 and 2.1 mM, respectively, as well as for (−)-epicatechin with K_m equal to 0.65 and 1.18 mM according to Liu et al. [41] and Martinez-Cayuela et al. [42], respectively. The kinetic constant for catechol was 0.61 mM, similar to the value range from 0.18 to 0.97 mM for cocoa pulp and bean, respectively [1], but different from other plant samples, with values ranging from 7.9×10^{-4} to 203.8 mM [2,28,43], which can be due to method of extraction, nature of the subtrate, and method of measurement.

Additionally, denatured PPO enzyme recovered using the optimized variables (70 mM inhibitory solution at 96 °C for 6.4 min) for the oxidation of catechol was also studied. Results showed that catechol had the lowest catalytic power, with 12.76 U mM^{-1} mL^{-1} min^{-1}; therefore, its affinity for the substrate (K_m) was reduced up to 13.7-fold. The lowest affinity of the enzyme could be a consequence of the degradation of the protein site of the enzyme and/or morphological changes in the enzyme, which can be related to its low protein content (16.04 ± 3.96 mg mL^{-1}) and low specific activity (2.12 ± 0.65 U$_{PPO}$ mg^{-1}) after the heat treatment, respectively.

Overall, these findings reinforce the high affinity of PPO for small *o*-diphenols, and thermal enzyme denaturation, and also highlight the importance of inhibiting PPO activity for controlling the dramatic loss of polyphenols by the enzymatic action.

3.5. Microscopy Analysis

An evaluation of the effect of PPO inhibition on morphology and cell wall was carried out by scanning electron microscopy. As can be seen in Figure 3A, fresh sample (non-treated sample) maintains the cell walls, which are well oval-shaped, non-fractured, solid and denser in appearance, and have the cellular content clearly embedded in them (fractal dimension of the images, FDt = 2.542). Figure 3B clearly showed evident changes after the PPO inhibition, which consisted of faster changes in temperature between 4–97 °C for short times (1–6 min), with significant differences ($p <$ 0.05) observed in the FDt values. In fact, the FDt is an important tool for image analysis that makes it possible to quantify the image texture with the aim of identifying significant differences between treatments. Moreover, the cell walls become larger, smooth, fibrous and unfolded, and show evidence of more interspace and holes in the microstructure (Figure 3B) as a result of the enzyme thermal denaturation or the effect of heat shock proteins (FDt = 2.872), which are consistent with Terefe et al. [44]. This observation reinforced that heat treatment is a faster and robust method to change the conformational structure of PPO enzyme, and thus reduce its activity in cocoa beans and increase the extraction yield of polyphenols.

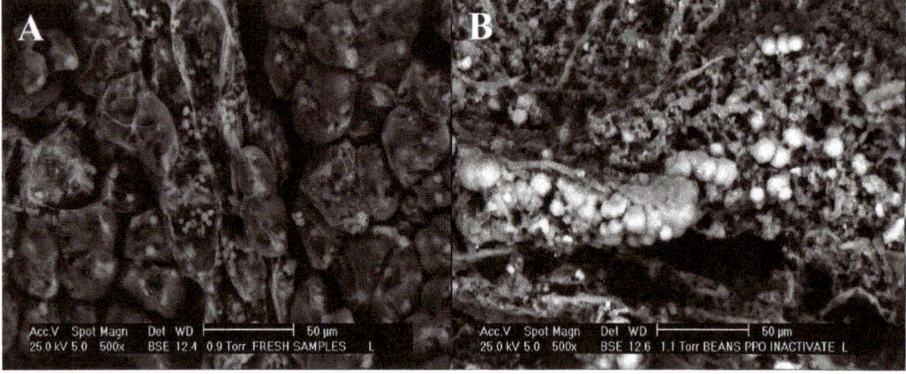

Figure 3. Microscopy images for the microstructure of (**A**) non-treated sample (fresh cocoa bean) and (**B**) sample after optimal conditions for the inhibition of PPO enzyme (70 mM inhibitory solution at 96 °C, 6.4 min).

4. Conclusions

In this study, experimental conditions to inhibit the action of PPO enzyme in terms of the specific activity of PPO measurement and total polyphenol content were optimized. Our study has reported, for the first time, an equation that correlates both high recovery of total polyphenols and high inhibition of PPO enzyme in cocoa beans. Confirmation of heat denaturation during the inactivation process by SEM images, and the high affinity of PPO for small o-diphenols, especially for catechol, followed by (−)-epicatechin, have also been studied. This work provides a promising, robust, easier, and food-safe procedure for obtaining enriched polyphenol extract with longer oxidative enzyme stability.

Supplementary Materials: The following are available online at http://www.mdpi.com/2076-3921/9/6/458/s1, Figure S1: Relationship between specific activity of PPO and total polyphenol content on cocoa beans. Figure S2: Correlation Γ (See Equation (5) as a function a) PPO inhibition (%), and (b) Total polyphenol (%) on cocoa beans.

Author Contributions: Conceptualization, S.T.-U., J.G.-C., A.R.V.-J., M.d.J.P.-F. and L.J.L.-G.; methodology, S.T.-U.; formal analysis S.T.-U., J.G.-C., M.d.J.P.-F. and L.J.L.-G.; investigation, S.T.-U., J.G.-C., M.d.J.P.-F. and L.J.L.-G.; writing—original draft preparation, S.T.-U.; writing—review and editing, S.T.-U. and L.J.L.-G. All authors have read and agreed to the published version of the manuscript.

Funding: This research was funded by Universidad Industrial de Santander, grant numbers 2341 and 2511, in addition to the Department of Science, Technology and Innovation (COLCIENCIAS-Colombia), grant number 567 of 2012.

Acknowledgments: Said Toro wants to thank COLCIENCIAS (Colombia) for a financial scholarship. Said also gratefully acknowledges the analysis support and insightful comments of María Ximena Quintanilla. The authors are grateful to FEDECACAO (especially Nubia Martinez) for their technical support and thank Vicerrectoría de Investigación y Extensión from Universidad Industrial de Santander for their financial support that made this work possible.

Conflicts of Interest: The authors declare no conflict of interest.

References

1. Macedo, A.S.L.; de Rocha, F.S.; da Ribeiro, M.S.; Soares, S.E.; da Bispo, E.S. Characterization of polyphenol oxidase in two cocoa (*Theobroma cacao* L.) cultivars produced in the south of Bahia, Brazil. *Food Sci. Technol.* **2016**, *36*, 56–63. [CrossRef]

2. Doğru, Y.Z.; Erat, M. Investigation of some kinetic properties of polyphenol oxidase from parsley (*Petroselinum crispum*, Apiaceae). *Food Res. Int.* **2012**, *49*, 411–415. [CrossRef]

3. De Oliveira Carvalho, J.; Orlanda, J.F.F. Heat stability and effect of pH on enzyme activity of polyphenol oxidase in buriti (*Mauritia flexuosa* Linnaeus f.) fruit extract. *Food Chem.* **2017**, *233*, 159–163. [CrossRef] [PubMed]

4. Li, F.; Chen, G.; Zhang, B.; Fu, X. Current applications and new opportunities for the thermal and non-thermal processing technologies to generate berry product or extracts with high nutraceutical contents. *Food Res. Int.* **2017**, *100*, 19–30. [CrossRef] [PubMed]

5. Osuga, D.; Van Der Schaaf, A.; Whitaker, J.R. Control of polyphenol oxidase activity using a catalytic mechanism. In *Protein Structure-Function Relationships in Foods*; Springer: Boston, MA, USA, 1994; pp. 62–88. ISBN 978-1-4615-2670-4.

6. Araji, S.; Grammer, T.A.; Gertzen, R.; Anderson, S.D.; Mikulic-Petkovsek, M.; Veberic, R.; Phu, M.L.; Solar, A.; Leslie, C.A.; Dandekar, A.M.; et al. Novel roles for the polyphenol oxidase enzyme in secondary metabolism and the regulation of cell death in walnut. *Plant Physiol.* **2014**, *164*, 1191–1203. [CrossRef] [PubMed]

7. Selamat, J.; Bakar, J.; Saari, N. Oxidation of polyphenols in unfermented and partly fermented cocoa beans by cocoa polyphenol oxidase and tyrosinase. *J. Sci. Food Agric.* **2002**, *82*, 559–566.

8. Cheema, S.; Sommerhalter, M. Characterization of polyphenol oxidase activity in Ataulfo mango. *Food Chem.* **2015**, *171*, 382–387. [CrossRef]

9. De Jesus, A.L.T.; Leite, T.S.; Cristianini, M. High isostatic pressure and thermal processing of açaí fruit (*Euterpe oleracea* Martius): Effect on pulp color and inactivation of peroxidase and polyphenol oxidase. *Food Res. Int.* **2018**, *105*, 853–862. [CrossRef]

10. De la Rosa, L.A.; Alvarez-Parrilla, E.; González-Aguilar, G.A. *Fruit and Vegetable Phytochemicals: Chemistry, Nutritional Value, and Stability*, 1st ed.; De la Rosa, L.A., Ed.; Wiley-Blackwell: Ames, Iowa, 2009; ISBN 9780813803203.

11. He, Q.; Luo, Y.; Chen, P. Elucidation of the mechanism of enzymatic browning inhibition by sodium chlorite. *Food Chem.* **2008**, *110*, 847–851. [CrossRef]

12. Mishra, B.B.; Gautam, S.; Sharma, A. Free phenolics and polyphenol oxidase (PPO): The factors affecting post-cut browning in eggplant (*Solanum melongena*). *Food Chem.* **2013**, *139*, 105–114. [CrossRef]

13. De Leonardis, A.; Angelico, R.; Macciola, V.; Ceglie, A. Effects of polyphenol enzymatic-oxidation on the oxidative stability of virgin olive oil. *Food Res. Int.* **2013**, *54*, 2001–2007. [CrossRef]

14. De Brito, E.S.; García, N.H.P.; Amâncio, A.C. Effect of polyphenol oxidase (PPO) and air treatments on total phenol and tannin content of cocoa nibs. *Food Sci. Technol.* **2002**, *22*, 45–48. [CrossRef]

15. Kruger, M.J.; Davies, N.; Myburgh, K.H.; Lecour, S. Proanthocyanidins, anthocyanins and cardiovascular diseases. *Food Res. Int.* **2014**, *59*, 41–52. [CrossRef]

16. Schinella, G.; Mosca, S.; Cienfuegos-Jovellanos, E.; Pasamar, M.Á.; Muguerza, B.; Ramón, D.; Ríos, J.L. Antioxidant properties of polyphenol-rich cocoa products industrially processed. *Food Res. Int.* **2010**, *43*, 1614–1623. [CrossRef]

17. Nakajima, V.M.; Moala, T.; Caria, C.R.e.P.; Moura, C.S.; Amaya-Farfan, J.; Gambero, A.; Macedo, G.A.; Macedo, J.A. Biotransformed citrus extract as a source of anti-inflammatory polyphenols: Effects in macrophages and adipocytes. *Food Res. Int.* **2017**, *97*, 37–44. [CrossRef]

18. Marchese, A.; Coppo, E.; Sobolev, A.P.; Rossi, D.; Mannina, L.; Daglia, M. Influence of in vitro simulated gastroduodenal digestion on the antibacterial activity, metabolic profiling and polyphenols content of green tea (*Camellia sinensis*). *Food Res. Int.* **2014**, *63*, 182–191. [CrossRef]

19. Karar, M.G.E.; Pletzer, D.; Jaiswal, R.; Weingart, H.; Kuhnert, N. Identification, characterization, isolation and activity against Escherichia coli of quince (*Cydonia oblonga*) fruit polyphenols. *Food Res. Int.* **2014**, *65*, 121–129. [CrossRef]

20. Alañón, M.E.; Castle, S.M.; Siswanto, P.J.; Cifuentes-Gómez, T.; Spencer, J.P.E. Assessment of flavanol stereoisomers and caffeine and theobromine content in commercial chocolates. *Food Chem.* **2016**, *208*, 177–184. [CrossRef] [PubMed]

21. Lamuela-Raventós, R.M.; Romero-Pérez, A.I.; Andrés-Lacueva, C.; Tornero, A. Review: Health effects of cocoa flavonoids. *Food Sci. Technol. Int.* **2005**, *11*, 159–176. [CrossRef]

22. Pedan, V.; Fischer, N.; Bernath, K.; Hühn, T.; Rohn, S. Determination of oligomeric proanthocyanidins and their antioxidant capacity from different chocolate manufacturing stages using the NP-HPLC-online-DPPH methodology. *Food Chem.* **2017**, *214*, 523–532. [CrossRef]

23. Guerrero-Beltrán, J.A.; Swanson, B.G.; Barbosa-Cánovas, G.V. Inhibition of polyphenoloxidase in mango puree with 4-hexylresorcinol, cysteine and ascorbic acid. *LWT Food Sci. Technol.* **2005**, *38*, 625–630. [CrossRef]

24. Pizzocaro, F.; Torreggiani, D.; Gilardi, G. Inhibition of apple polyphenol oxidase (ppo) by ascorbic acid, citric acid and sodium chloride. *J. Food Process. Preserv.* **1993**, *17*, 21–30. [CrossRef]

25. Menon, A.S.; Hii, C.L.; Law, C.L.; Suzannah, S.; Djaeni, M. Effects of water blanching on polyphenol reaction kinetics and quality of cocoa beans. In *AIP Conference Proceedings*; AIP Publishing LLC: Melville, NY, USA, 2015; Volume 1699, pp. 030006–1–030006–7.

26. Ali, H.M.; El-Gizawy, A.M.; El-Bassiouny, R.E.I.; Saleh, M.A. Browning inhibition mechanisms by cysteine, ascorbic acid and citric acid, and identifying PPO-catechol-cysteine reaction products. *J. Food Sci. Technol.* **2015**, *52*, 3651–3659. [CrossRef] [PubMed]

27. Lee, C.Y.; Whitaker, J.R. (Eds.) *Enzymatic Browning and Its Prevention*, 1st ed.; ACS Symposium Series; American Chemical Society: Washington, DC, USA, 1995; Volume 600, ISBN 0-8412-3249-0.

28. Siddiq, M.; Dolan, K.D. Characterization and heat inactivation kinetics of polyphenol oxidase from blueberry (*Vaccinium corymbosum* L.). *Food Chem.* **2017**, *218*, 216–220. [CrossRef]

29. Pons-Andreu, J.-V.; Cienfuegos-Jovellanos, E.; Ibarra, A. Process for Producing Cocoa Polyphenol Concentrate. U.S. Patent 0193629 A1, 2008.

30. Manzocco, L.; Anese, M.; Nicoli, M.C. Radiofrequency inactivation of oxidative food enzymes in model systems and apple derivatives. *Food Res. Int.* **2008**, *41*, 1044–1049. [CrossRef]

31. Yuan, G.; Lv, H.; Tang, W.; Zhang, X.; Sun, H. Effect of chitosan coating combined with pomegranate peel extract on the quality of Pacific white shrimp during iced storage. *Food Control* **2016**, *59*, 818–823. [CrossRef]

32. Babu, B.R.; Rastogi, N.K.; Raghavarao, K.S.M.S. Liquid-liquid extraction of bromelain and polyphenol oxidase using aqueous two-phase system. *Chem. Eng. Process. Process Intensif.* **2008**, *47*, 83–89. [CrossRef]
33. Bradford, M.M. A rapid and sensitive method for the quantitation of microgram quantities of protein utilizing the principle of protein-dye binding. *Anal. Biochem.* **1976**, *72*, 248–254. [CrossRef]
34. AOCS. *Official Methods of Analysis of AOAC International*; AOCS: Washington, DC, USA, 1990; ISBN 0935584544.
35. Singleton, V.L.; Orthofer, R.; Lamuela-Raventós, R.M. Analysis of total phenols and other oxidation substrates and antioxidants by means of folin-ciocalteu reagent. *Methods Enzymol.* **1998**, *299*, 152–178.
36. Hernández-Carrión, M.; Hernando, I.; Sotelo-Díaz, I.; Quintanilla-Carvajal, M.X.; Quiles, A. Use of image analysis to evaluate the effect of high hydrostatic pressure and pasteurization as preservation treatments on the microstructure of red sweet pepper. *Innov. Food Sci. Emerg. Technol.* **2015**, *27*, 69–78. [CrossRef]
37. Lee, P.M.; Lee, K.H.; Karim, M.I.A. Biochemical-Studies of Cocoa Bean Polyphenol Oxidase. *J. Sci. Food Agric.* **1991**, *55*, 251–260. [CrossRef]
38. Dudley, E.D.; Hotchkiss, J.M. Cysteine as an inhibitor of polyphenoloxidase. *J. Food Biochem.* **1989**, *13*, 65–75. [CrossRef]
39. Wuyts, N.; De Waele, D.; Swennen, R. Extraction and partial characterization of polyphenol oxidase from banana (*Musa acuminata* Grande naine) roots. *Plant Physiol. Biochem.* **2006**, *44*, 308–314. [CrossRef] [PubMed]
40. Ho, K.K. Characterization of polyphenol oxidase from aerial roots of an orchid, Aranda "Christine 130". *Plant Physiol. Biochem.* **1999**, *37*, 841–848. [CrossRef]
41. Liu, L.; Cao, S.; Xie, B.; Sun, Z.; Li, X.; Miao, W. Characterization of polyphenol oxidase from litchi pericarp using (−)-epicatechin as substrate. *J. Agric. Food Chem.* **2007**, *55*, 7140–7143. [CrossRef]
42. Martinez-Cayuela, M.; De Medina, L.S.; Faus, M.J.; GilL, A. Cherimoya (*Annona cherimola* Mill) polyphenoloxidase: Monophenolase and dihydroxyphenolase activities. *J. Food Sci.* **1988**, *53*, 1191–1194. [CrossRef]
43. Dincer, B.; Colak, A.; Aydin, N.; Kadioglu, A.; Güner, S. Characterization of polyphenoloxidase from medlar fruits (*Mespilus germanica* L., Rosaceae). *Food Chem.* **2002**, *77*, 1–7. [CrossRef]
44. Terefe, N.S.; Delon, A.; Buckow, R.; Versteeg, C. Blueberry polyphenol oxidase: Characterization and the kinetics of thermal and high pressure activation and inactivation. *Food Chem.* **2015**, *188*, 193–200. [CrossRef]

 *antioxidants*

Article

Flavanol Polymerization Is a Superior Predictor of α-Glucosidase Inhibitory Activity Compared to Flavanol or Total Polyphenol Concentrations in Cocoas Prepared by Variations in Controlled Fermentation and Roasting of the Same Raw Cocoa Beans

Kathryn C. Racine [1], Brian D. Wiersema [1], Laura E. Griffin [2], Lauren A. Essenmacher [1], Andrew H. Lee [1], Helene Hopfer [3], Joshua D. Lambert [3], Amanda C. Stewart [1] and Andrew P. Neilson [2,*]

[1] Department of Food Science and Technology, Virginia Polytechnic Institute and State University, Blacksburg, VA 24060, USA; kcracine@ncsu.edu (K.C.R.) wiersema@vt.edu (B.D.W.); elauren7@vt.edu (L.A.E.); andhlee@vt.edu (A.H.L.); amanda.stewart@vt.edu (A.C.S.)

[2] Plants for Human Health Institute, Department of Food, Bioprocessing and Nutrition Sciences, North Carolina State University, Kannapolis, NC 28081, USA; legriff2@ncsu.edu or legriffin18@gmail.com

[3] Department of Food Science, Pennsylvania State University, University Park, PA 16801, USA; hxh83@psu.edu (H.H.); jdl134@psu.edu (J.D.L.)

* Correspondence: aneilso@ncsu.edu; Tel.: +1-704-250-5495; Fax: +1-704-250-5409

Received: 13 November 2019; Accepted: 8 December 2019; Published: 11 December 2019

Abstract: Raw cocoa beans were processed to produce cocoa powders with different combinations of fermentation (unfermented, cool, or hot) and roasting (not roasted, cool, or hot). Cocoa powder extracts were characterized and assessed for α-glucosidase inhibitory activity in vitro. Cocoa processing (fermentation/roasting) contributed to significant losses of native flavanols. All of the treatments dose-dependently inhibited α-glucosidase activity, with cool fermented/cool roasted powder exhibiting the greatest potency (IC_{50}: 68.09 μg/mL), when compared to acarbose (IC_{50}: 133.22 μg/mL). A strong negative correlation was observed between flavanol mDP and IC_{50}, suggesting flavanol polymerization as a marker of enhanced α-glucosidase inhibition in cocoa. Our data demonstrate that cocoa powders are potent inhibitors of α-glucosidase. Significant reductions in the total polyphenol and flavanol concentrations induced by processing do not necessarily dictate a reduced capacity for α-glucosidase inhibition, but rather these steps can enhance cocoa bioactivity. Non-traditional compositional markers may be better predictors of enzyme inhibitory activity than cocoa native flavanols.

Keywords: flavan-3-ol; procyanidin; α-glucosidase; melanoidin; Maillard reaction

---

1. Introduction

Cocoa beans (*Theobroma cacao*) are concentrated dietary sources of flavanols, a subclass of polyphenols, which are thought to be responsible for many of the bioactivities of cocoa [1–5]. (−)-epicatechin and (+)-catechin (which is epimerized to (−)-catechin during roasting) are the major monomeric flavanols in raw cocoa. Native flavanols in cocoa beans are approximately 58% procyanidins (PCs), or flavanol oligomers and polymers (monomeric residues linked via 4→β6 or 4→β8 bonds). These flavanols exhibit potent antioxidant and health-protective activities, including the modulation of oxidative stress and potentially reducing the risk of various chronic conditions, such as cardiovascular disease (CVD), type II diabetes mellitus (T2D), and different forms of cancer. Additionally, cocoa beans contain other bioactives, such as lipids, fiber, lignins, melanoidins (after roasting), methylxanthines, and other complex compounds that have not been extensively characterized. Beans contain approximately 55% fat, 16% fiber, 10% protein, and 3% ash, depending on variety. The health benefits associated with dietary cocoa are likely due to multiple bioactive compounds and their interactions, rather than one compound or class of compounds, because of the complex composition of cocoa and the reactions that occur during cocoa processing [6].

After harvesting, cocoa beans undergo a series of processing steps, including fermentation, drying, roasting, winnowing, and various other processes that may include pressing or alkalization, to produce a final product, such as cocoa powder or chocolate. These processes strongly influence the chemical composition of the product, with fermentation resulting in approximately 0–70% loss of total polyphenols and roasting costing an additional 15–40% loss [7–9]. Additionally, non-enzymatic browning reactions occur between native cocoa polyphenols and mono- or polysaccharides, proteins, and amino acids to produce Maillard reaction products, most notably melanoidins [10]. The widely-accepted paradigm is that preservation of native flavanols is critical for retaining bioactivity [7,9,11]. However, it is possible that reactions occurring throughout processing may generate processing-derived compounds with novel activities, such as lignin-like complexes and melanoidins, potentially preserving or even enhancing certain bioactivities as compared to the raw cocoa bean [7,12–19]. The levels and activities of these large, complex, and diverse compounds in cocoa are largely unknown due to analytical challenges (such as lack of authentic standards, structural complexity and size of potential products), low bioavailability of large complex compounds, and the continued research focus on small monomeric flavanols.

In vivo, flavanols have highly variable systemic oral bioavailability, with absorption being inversely proportional to molecular weight and ~0% absorption for compounds $\geq$ tetramer [20–24]. Thus, the actions of these bioactives are likely to predominantly occur in the lumen and epithelium of the gastrointestinal (GI) tract, where the delivery of native flavanols and processing products is high and not subject to various barriers and metabolism/transport processes that reduce flavanol concentration and activity. Bioactivity exerted within the lumen and epithelium of the gut might play a key role in the mitigation and prevention of obesity and related conditions, such as T2D and CVD. Specifically, the inhibition of gut digestive enzymes to limit macronutrient digestion is a promising mode of bioactivity that does not require systemic bioavailability. Previous studies demonstrate that cocoa flavanols inhibit lipases, α-amylase, and α-glucosidase; α-glucosidase appears to be the most susceptible to cocoa inhibition [13,25]. α-glucosidase is a brush border enzyme that hydrolyzes starch and maltose into absorbable glucose [26]. The inhibition of α-glucosidase is a potential strategy for inhibiting or slowing blood glucose absorption in the context of glucose intolerance. Commercially available α-glucosidase inhibitors, such as acarbose, miglitol, and voglibose, come with high prices and various side effects (such as GI discomfort), therefore warranting the investigation of dietary flavonoids (from berries, red wine, green tea, cocoa, etc.) as potential inhibitors [16]. While isolated PCs are effective α-glucosidase inhibitors, we recently found that various cocoas were effective inhibitors of α-glucosidase, despite large reductions in native flavanols as a result of fermentation and roasting [13]. These results challenge the idea that losses of native flavanols inherently reduces all bioactivity of cocoa and, therefore, warrant further investigation into the factors that determine α-glucosidase inhibition. Specifically, the impact of cocoa processing on subsequent bioactivity and the identification of non-flavanol cocoa components with bioactivity are of interest.

Fermentation and roasting are the logical steps in cocoa processing to manipulate in order to produce a final product with variable flavanol composition. Sourcing cocoas with the same bean origin and known processing history is impractical if not impossible, due to poor bean traceability, poor documentation, and varying practices of fermentation. A controlled model pilot-scale fermentation using a common starting material is necessary to conduct research regarding the impact of fermentation and further processing on bean composition and subsequent bioactivity [27–29]. Therefore, the main objectives of this work were to (1) evaluate the effect of extremes in fermentation and roasting on the composition of cocoa beans and powder, (2) determine how fermentation and roasting affect α-glucosidase inhibitory activity of cocoa powder, and (3) identify the compositional factors and processing conditions that optimize α-glucosidase inhibitory activity of cocoa.

2. Materials and Methods

2.1. Chemicals and Standards

Cargill, Inc. (Wayzata, MN, USA) generously provided raw unfermented cocoa beans, sourced from Hispaniola. The beans were stored in burlap sacks at 3.5 °C prior to use. LC-MS grade acetonitrile (ACN), LC-MS grade methanol (MeOH), citric acid, yeast extract, malt extract, calcium-lactate pentahydride, tween 80, sodium hydroxide, magnesium sulfate heptahydride, manganese sulfate monohydride, sucrose, glucose, fructose, calcium carbonate, agar, and water were obtained from Thermo-Fisher Scientific (Waltham, MA, USA). Glacial acetic acid, methanol, and acetone were obtained from VWR (Radnor, PA, USA). The standards of (−)-epicatechin (EC), (±)-catechin (C), and procyanidin B2 (PCB2) were obtained from ChromaDex (Irving, CA). Standards of procyanidin C1 (PCC1), cinnamtannin A2 (CinA2), and DP 5-9 purified from cocoa (purity: DP 3-5: 93–99%, DP6-9: 80–92%) were obtained from Planta Analytica (New Milford, CT, USA). Ammonium formate, Folin-Ciocalteu reagent, 4-dimethylaminocinnamaldehyde (DMAC), and α-glucosidase (from *Saccharomyces cerevisiae*) were obtained from Sigma-Aldrich (G5003, St. Louis, MO, USA). The solvents were ACS grade or higher.

2.2. Fermentation Model System and Processing

A partial factorial approach (Figure 1A) was employed to generate cocoa powders from the same batch of raw beans: fermentation (unfermented (UF), cool fermented (CF), and hot fermented (HF)) and roasting (unroasted (UR), cool temperature roasted (CR), and hot temperature roasted (HR)). Certain possible combinations were not evaluated due to cost constraints.

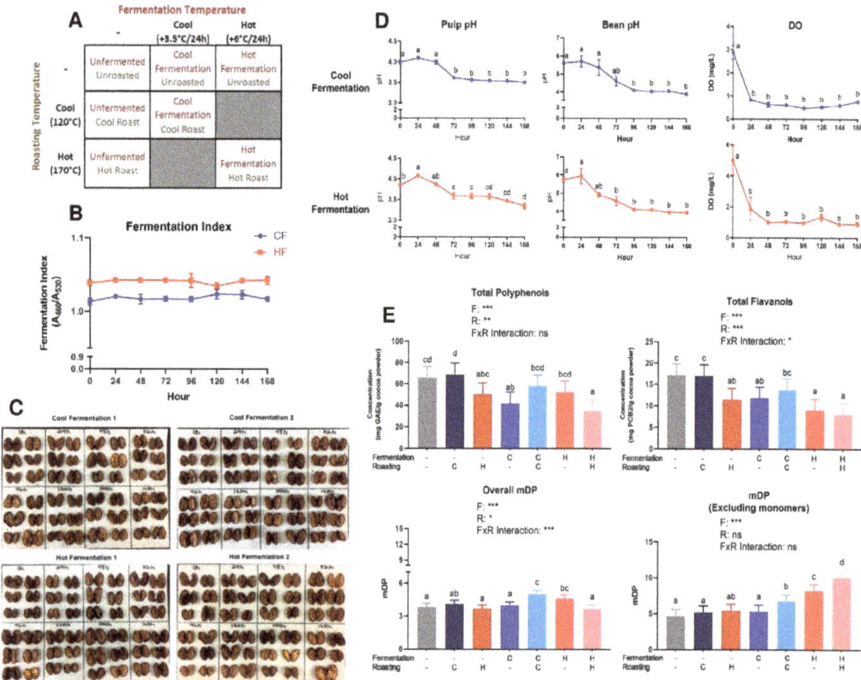

Figure 1. (**A**) Cocoa processing model system to evaluate the impact of combined fermentation and roasting parameters, producing seven total cocoa powders: unfermented/unroasted (UF/UR), unfermented/cool roast (UF/CR), unfermented/hot roast (UF/HR), cool fermentation/unroasted (CF/UR), cool fermentation/cool roast (CF/CR), hot fermentation/unroasted (HF/UR), and hot fermentation/hot roast (HF/HR). (**B**) Fermentation index as a ratio of absorbance at 460 nm:530 nm, with values ≥1 indicating a complete fermentation. Note broken axes for ease of interpretation on select graphs. Values are presented as the mean ± SEM of fermentation replicates within treatments. Significant between time points for each value was determined by one-way ANOVA and Tukey's HSD post-hoc test ($p < 0.05$). Time points with different letters are significantly different within values. (**C**) Cut test for all fermentations performed. (**D**) pH of simulated pulp media/bean nib and dissolved oxygen (DO). It is important to note that for bean nib measurements, these values do not quantify the pH of the cocoa bean itself, but rather the acidity derived from bean acids diluted in water. These nib values are useful for comparison between the pH of the solution produced by beans at different time points. (**E**) Total polyphenols in each cocoa powder, expressed in gallic acid equivalents. Total flavanols from cocoa powder expressed in procyanidin B2 equivalents. Overall mean flavanol degree of polymerization (mDP) for the total flavanols in cocoa powder- native monomers were accounted for in calculation. Mean flavanol degree of polymerization for oligomers and polymers in cocoa powder (not including native monomers); Note broken axes for ease of interpretation. All values are presented as the least squares (LS) means with upper and lower confidence interval (CI). Significance between treatments was determined by two-way ANOVA for the roasting and fermentation temperature effects using type III sums of squares to account for unbalanced data, followed by post-hoc comparisons of LSMEANS ($p < 0.05$). Normality was checked for each variable visually and with the Shapiro–Wilks test, and if needed, transformed (Log or Box–Cox) prior to running the ANOVA and post-hoc test. Treatments with different superscript letters (a–d) in Figure 1D and E are significantly different within values. Legends above graphs indicate treatment (F = fermentation; R = roasting) main effect and interactions as determined by two-way ANOVA, * $p < 0.05$, ** $p < 0,01$, *** $p < 0.001$.

2.2.1. Fermentation

The raw unfermented cocoa beans were rehydrated, fermented, and dried based on previously established pilot-scale cocoa fermentation protocols described by Racine et al. [30] and Lee et al. [31] with modifications. The unfermented cocoa beans (32 kg) were rehydrated in plastic fermentation boxes (polypropylene, Sterilite, Townsend, MA, USA) by submersion in distilled, deionized (DI) water (45 L) for 24 h at room temperature. The final moisture content of the beans after rehydration was between 35–50% (IR-120 Moisture Analyzer, Denver Instrument, Bohemia, NY, USA). After draining off the excess water, rehydrated beans (60 kg) were mixed with 60 L of simulated pulp media (Table 1) (three replicate fermentation boxes per fermentation run; approximately 20 kg rehydrated beans; and, 20 L simulated pulp media per box). Boxes were covered and placed inside a pre-heated (25 °C) incubator (Forma 29 cu ft Reach-In-Incubator, Model No. 3950, Thermo Fisher Scientific, Waltham, MA, USA).

Table 1. Composition of simulated pulp media.

Reagent [e]	Mass [a] (g)
Simulated Pulp Solution [b]	
Citric acid	40
Yeast extract	20
Peptone	20
Calcium lactate pentahydrate	4
Tween 80	4
Magnesium-Manganese Solution [c]	
Magnesium sulfate-heptahydride	2
Manganese sulfate-monohydride	0.8
Sugar Solutions [d]	
Sucrose	100
Glucose	160
Fructose	180

[a] per four liters of solution; [b] reagents were combined with 1600 mL DI water, pH adjusted to 3.6 using 1 N NaOH, and adjusted to a final volume of 2.4 L DI water before autoclaving; [c] reagents were combined with 400 mL of DI water; [d] reagents were made separately into 400 mL solutions; [e] solutions c and d were combined with autoclaved solution b to begin fermentation.

In total, four fermentation runs were conducted: two cool and two hot fermentation runs. Each fermentation run continued for a total of 168 h. For each run, the material was fermented in three separate boxes within the same incubator, with 20 kg rehydrated beans and 20 L simulated pulp in each box. The incubator set point was raised 3.5 °C/24 h for the cool fermentation runs and 6 °C/24 h for the hot fermentation runs, to final temperatures of 46°C (cool) and 60°C (hot) (Figure 1A). For all of the boxes in all fermentation runs, the beans were manually agitated for 5 min. every 8 h to ensure that the simulated pulp media was well-mixed and properly aerated. The pulp dissolved oxygen (DO) and pH values were monitored while using benchtop meters (Orion DO Probe 083005MD; Orion Versa Star Pro pH meter; Thermo Fisher Scientific, Waltham, MA, USA). Figure S1 shows images of a representative fermentation batch over time. After 168 h of fermentation, the beans were drained to remove the remaining simulated pulp media, evenly spread onto baking sheets, and then oven dried (Rational, Germany; Blodgett, Burlington, VT, USA) at 65.5 °C for 24–26 h or until the moisture content fell below 8%. After drying, all of the beans from both cool runs were thoroughly commingled and stored at 3.5 °C until roasting, and the same was done for all beans from both hot runs. The beans that were subjected to the unfermented treatment were immediately oven dried following rehydration and draining, according to the procedures above.

2.2.2. Roasting and Further Processing

Roasting was performed in a gas-fired drum roaster (180 kg capacity, Probat, Inc., Vernon Hills, IL) at a drum speed of 15 rpm in collaboration with Epiphany Craft Malt (Durham, NC, USA). Each unique fermentation/roasting treatment was separately roasted in batches of approximately 30 kg. The cool roasted treatment temperature was 120 °C and the hot roasted treatment temperature was 170 °C (Figure 1A). Roasting treatments were conducted for 20 min. each. After roasting, the beans were placed on a rotating cooling table and then stored at 3.5 °C until further processing. The beans were further processed into cocoa powder in collaboration with Blommer Chocolate Company (East Greenville, PA, USA). The beans were first winnowed to remove shells and ground into liquor. The liquor was then heated to approximately 200 °C and pressed (Cacao Cucino, Model No. 306487, Clearwater, FL, USA) for 133 min. to produce a solid cake that was then ground into a homogenous cocoa powder. All seven treatments were uniformly processed into cocoa powder. The powders were stored at −20 °C until further analysis. The moisture and fat content for liquors and cake (ORACLE Rapid Fat Analyzer, CEM, Matthews, NC, USA) and particle size of liquors (Microtrac S3500, Microtrac, Montgomeryville, PA, USA) was measured for each treatment per Blommer standard operating procedures.

2.3. Fermentation Assays

The bean pH was determined every 24 h based on the method that was described by Racine et al. [30]. A representative cut test was performed on a sample of whole beans from each of the four fermentation runs (2 cool batches, 2 hot batches). Beans ($n = 6$) from each 24 h sampling period (0–168 h) were cut through the middle lengthwise so that color and quality could be assessed. This test is a standard assessment of post-fermentation bean quality and suitability to move forward in processing [32,33]. The cocoa bean fermentation index (FI) was measured every 24 h during fermentation based on the method of Romero-Cortes et al. [34] with minor modifications. Randomly selected whole cocoa beans ($n = 3$–5) were frozen with liquid nitrogen and ground into a powder in an electric spice grinder. The powder (50 mg) was mixed with 5 mL MeOH/HCl (97:3 v/v) and extracted at 4 °C for 16–18 h on a rotating platform. The samples were then centrifuged (5 min., 3500× g), supernatant collected (300 μL), and absorbance measured at 460 nm and 530 nm using a 96-well microplate. The FI was calculated while using the equation below and all analyses were performed in analytical triplicate.

$$FI = \frac{A_{460}}{A_{530}} \tag{1}$$

2.4. Polyphenol Extraction and Characterization

Polyphenol-rich extracts were made from raw beans, the fermented cocoa beans, and powders for all seven treatments, as described previously [13,30]. The Folin–Ciocalteu colorimetric assay was used to approximate the total phenolic content of the freeze-dried cocoa extracts, and total flavanols were measured by the 4-dimethylaminocinnamaldehyde (DMAC) colorimetric assay, as previously described in Dorenkott et al. [12]. These values were expressed in mg Gallic Acid Equivalents (mg GAE)/g bean and mg PCB2/g bean, respectively. The mean degree of polymerization (mDP) of the flavanols was determined by thiolysis based on the protocol of Dorenkott et al. [12], with minor modifications. Cocoa monomeric flavanols and PCs (DP 1-10) were quantified by HILIC UPLC-MS/MS based on the method of Racine et al. [35]. Refer to Supplementary Materials, including Table S1, for full methodological details.

2.5. Melanoidin Dialysis

The dialysis method that was proposed by Sacchetti et al. [36] was followed with modifications to isolate high and low molecular weight (HMW and LMW, respectively) extract fractions. Polyphenol-rich extracts were re-dissolved in extraction solution (70:28:2 acetone, water, acetic acid) at 40 mg/mL. Dialysis was performed in triplicate while using acidified methanol: water (60:40, 0.1% formic acid) as the dialysis solvent. For each replicate, 2.5 mL of dissolved extract (i.e. 100 mg) was placed inside presoaked dialysis tubing (8–10 kDa MW cutoff, Spectrum Spectra/Por Biotech-Grade RC Dialysis, Fisher) and clipped closed. This MW cutoff was chosen based on preliminary data (Table S2) illustrating that the majority of early, intermediate, and final Maillard reaction products (MRP) are eluted from the 3.5–5 kDa and 8–10 kDa tubing, with very little in 20 kDa, followed by a significant increase in 50 kDa (see Supplementary Materials). The tubing was submerged in 250 mL dialysis solvent and the beaker was stirred at 4 °C for 24 h and continuously sparged with N_2. Following dialysis, the solvent outside the tubing and the components remaining within the tubing were separately rotary evaporated at 55 °C, frozen at −80°C, and then freeze-dried. Following freeze-drying, the solids were weighed, and yield was calculated.

Fractions and undialyzed extracts were resolubilized in 0.05 M H_2SO_4 to a concentration of 0.15625 mg/mL (for detection of early MRPs in all samples), 2.5 mg/mL (for detection of intermediate MRPs in all samples), 5 mg/mL (for detection of late MRPs in LMW, <8–10 kDa, fractions), 2.5 mg/mL (for detection of late MRPs in HMW, >8–10 kDa, fractions), and 10 mg/mL (for detection of late MRPs in unfractionated extracts) to selectively quantify melanoidins/MRPs. Each diluted dialysate, non-dialyzable extract, and unfractionated extract was transferred (300 µL) into a UV-Star 96-well plate. The absorbance was read at 280 nm (early MRP), 360 nm (intermediate MRP), and 420 nm (late MRP). The MRP values are reported as absolute absorbance values (single dilution used for each fraction to facilitate direct comparisons of absorbance values: 5 mg/mL (<8–10 kDa), 2.5 mg/mL (>8–10 kDa), and 10 mg/mL (unfractionated extract)) due to the lack of a good analytical standard.

2.6. In Vitro α-Glucosidase Inhibition Assay

The powder extracts were screened for α-glucosidase inhibitory activity *in vitro*, as described previously [13]. A 0.1 M phosphate buffer (pH 6.9) was prepared in water with sodium phosphate monobasic anhydrous (8.05 g/L), and sodium phosphate dibasic anhydrous (4.67 g/L). pH was adjusted to 6.9 with 1 N NaOH. The extracts were diluted to 0.3125–8000 µg/mL in 10% DMSO. A negative control (no inhibitor, i.e. 0 µg/mL extract) was prepared with only 10% DMSO. The α-glucosidase solution (1 U/mL) and *p*-nitrophenyl α-D-glucopyranoside (*p*NPG, 1 mM) substrate solution were prepared in 0.1 M phosphate buffer. In 96-well plates (*n* = 6 wells/treatment), 50 µL of each working sample solution or negative control was mixed with 100 µL of α-glucosidase solution. The plate was then incubated at room temperature for 10 min., followed by the addition of 50 µL of *p*NPG solution to each well. Thus, the final concentration range of each inhibitor solution was 0–2000 µg/mL. The plate was then read at 405 nm. The negative control was 0 µg/mL extract and the positive control was acarbose in 10% DMSO (0–2000 µg/mL final concentration). The cocoa extracts were compared to the controls at each concentration and the values were expressed as % α-glucosidase activity while using the following equation:

$$\% \ \alpha - Glucosidase \ Activity = \left(\frac{\Delta A_{sample}}{\Delta A_{negative \ control}} \right) \times 100 \tag{2}$$

where:

ΔA_{sample}= the change in the individual absorbance value of the product of the inhibitor, substrate, and enzyme at each inhibitor dose before and after incubation; and,

$\Delta A_{negative \ control}$= the average change in absorbance of the negative control (0 µg/mL) before and after incubation.

α-glucosidase enzyme inhibition by all extracts and the positive control were assessed at physiologically relevant doses. The highest concentration (2000 µg extract/mL) is equivalent to approximately 13,333 µg powder/mL in the intestinal lumen (i.e. 13,333 ppm), when accounting for 15% extraction yield from powder. At an estimated 2 L upper digestive volume, this would equal approximately 26.67 g of cocoa product, or just less than 1 square of baking chocolate (1 square $\approx$ 28 g). At a lower concentration such as 100 µg/mL (666.67 µg original product/mL in the intestinal lumen), this would be equivalent to 1.33 g of original cocoa product or approximately 0.0475 squares of baking chocolate, which is a very achievable and physiologically relevant amount. Alternatively, acarbose is typically administered at 50–200 mg per dose [37]. At a 2 L upper digestive volume these would translate to 25–100 µg/mL, a range assessed in this assay and further confirming the physiological relevance of our concentrations.

2.7. Data Analysis and Statistics

All of the fermentation data (pH, FI, DO) within each treatment were analyzed by one-way ANOVA and when overall significance was detected, Tukey's HSD post hoc test was performed to determine the differences between time point means. Compositional data for beans and powders were analyzed for significance of main effect and interactions by two-way ANOVA using type III sums of squares to account for the unbalanced data (due to the partial factorial design) and if overall significance between treatments was detected, post-hoc comparisons of least-squares means (lsmeans) was performed. The normality was visually checked for each variable and with the Shapiro–Wilk test, and, if needed, transformed via Log or Box–Cox prior to two-way ANOVA and post-hoc test. Median inhibitory concentration (IC_{50}) values for α-glucosidase were calculated for each extract treatment while using a four-parameter sigmodal analysis. Enzyme inhibitory data were analyzed using a four-parameter log-logistic model for each of the seven treatments and the positive control, acarbose. Simple linear regression analysis was performed to correlate individual measured compositional factors (predictors: X) to α-glucosidase activity (IC_{50}, outcomes: Y). The mean compositional values and IC_{50} for each extract treatment were plotted (seven points per analysis), and R and R^2 were calculated. Statistical significance for all of the treatments in this study was defined *a priori* as $p < 0.05$. All of the analyses were carried out on GraphPad Prism v7.03 (GraphPad, La Jolla, CA, USA), R (v3.5.2) with Rstudio (v1.1.463, Boston, MA, USA) and additional packages, including lsmeans, car, RVAideMemoire, and others, to assist in the dose-response analyses.

3. Results

3.1. Fermentation Model System and Cocoa Processing

Monitored fermentation parameters (pH, DO) similarly progressed across 168 h for CF and HF, with initial (0 h) pulp pH values at 3.99 ± 0.05 and 3.84 ± 0.01, respectively, and ending (168 h) values at 3.51 ± 0.02 and 3.36 ± 0.07, respectively (Figure 1D). Furthermore, the bean pH values ranged from an initial 5.61 ± 0.01 (CF) to 5.75 ± 0.16 (HF) and concluded at 3.89 ± 0.04 (CF) and 3.95 ± 0.01 (HF) after 168 h. DO remained ≤1 mg/L after the first 24 h of fermentation. The initial FI values (Figure 1B) for both CF and HF were above 1.0 and remained within ±0.01 throughout the entire 168 h fermentation. Cut test results showed similar results, as there was no true progression of color from purple to brown beans. However, the sample size used was much smaller than what is traditionally seen, with the typical cut tests consisting of over 300 cut beans.

Table S3 reports the fat and moisture content of liquors and presscakes, as well as the particle size of liquors. While liquor fat content was generally similar across all treatments (~56–58%), UF/HR had much higher cake fat content (15.4%) and lower cake moisture (2.0%) than the other treatments. HF/HR had the lowest cake fat content, at 7.50%.

3.2. Characterization and Quantification of Polyphenols

Folin–Ciocalteu and DMAC, respectively, measured the total polyphenols and flavanols in beans (Figure S2) and powders (Figure 1E) from each treatment. The powders had a 2–3-fold higher total polyphenol and flavanol enrichment as compared to beans due to the concentration of polyphenols during pressing as cocoa butter is removed. Overall, treatment significantly influenced the measured total polyphenols and total flavanols (Figure 1E). Treatments with less harsh (i.e. cooler) processing conditions (UF/UR, UF/CR, CF/CR) generally had higher levels of total compounds than those that endured a more heat intensive processing (HF/UR, HF/HR, UF/HR), as expected, with the notable exception of CF/UR, which did not fit the overall pattern.

PCs (DP 1–10) were quantified from beans and powders from each of the seven treatments via HILIC UPLC-MS/MS. The powder concentrations were normalized to the fat-free mass in each treatment to account for the effect of pressing (i.e., fat content) and to present values only influenced by fermentation and roasting (Figure 2). Figures S3 and S4 show normalized values for powders and beans. Table S4 shows 2-way ANOVA results for monomers and PCs. As expected based on Folin–Ciocalteu and DMAC, UF/UR, and UF/CR powders had the highest concentrations of individual PCs DP 1–10 (Figure 2) and HF/HR powder had the lowest levels. However, HF/HR had the highest mDP of all treatments, at approximately 10, when calculated only factoring in oligomeric and polymeric PCs (Figure 1E), but this increase was not seen for mDP, including monomers present before thiolysis.

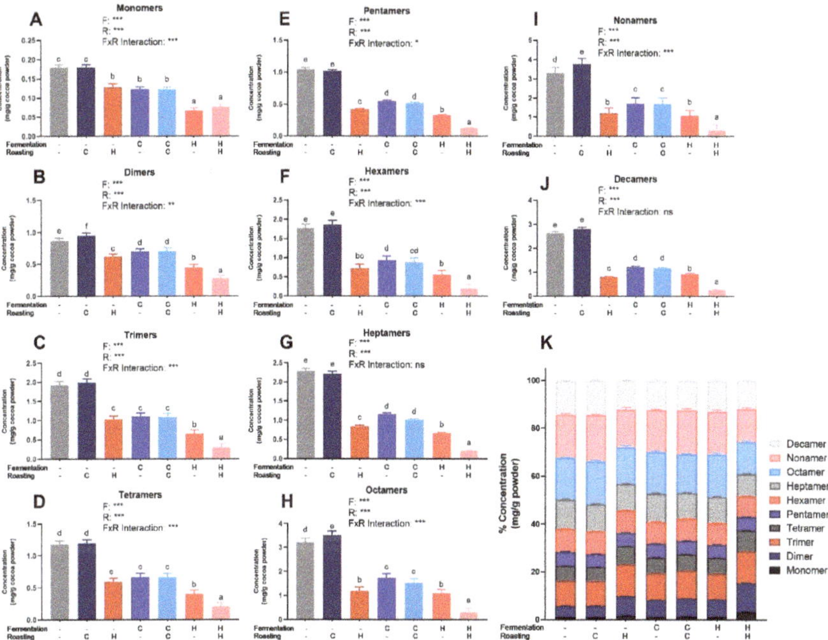

Figure 2. (**A–J**) Levels of individual procyanidin compounds in cocoa powders as measured by HILIC UPLC-MS/MS. (**K**) Individual values as a% of total measured procyanidins for each treatment. Values are normalized to the fat-free mass of each treatment to account to varying fat content. All values are presented as the LS means with upper and lower CI. Significance between treatments was determined by two-way ANOVA for the roasting and fermentation temperature effects using type III sums of squares to account for unbalanced data, followed by post-hoc comparisons of LSMEANS ($p < 0.05$). Normality was checked for each variable visually and with the Shapiro–Wilks test, and if needed, transformed (Log or Box-Cox) prior to running the ANOVA and post-hoc test. Treatments with different superscript letters (a–e) are significantly different within values. Legends above graphs indicate treatment (F = fermentation; R = roasting) main effect and interactions as determined by two-way ANOVA, * $p < 0.05$, ** $p < 0,01$, *** $p < 0.001$, refer to Table S4 for numerical values.

3.3. Quantification of Maillard Reaction Products

The MRPs were quantified in LMW and HMW fractions and unfractionated extracts (Figure 3). The results for early MRPs did not show clear trends. Intermediate and late MRPs generally increased as expected from UR to CR to HR within all of the treatments. Fermentation had little impact on MRPs, as expected.

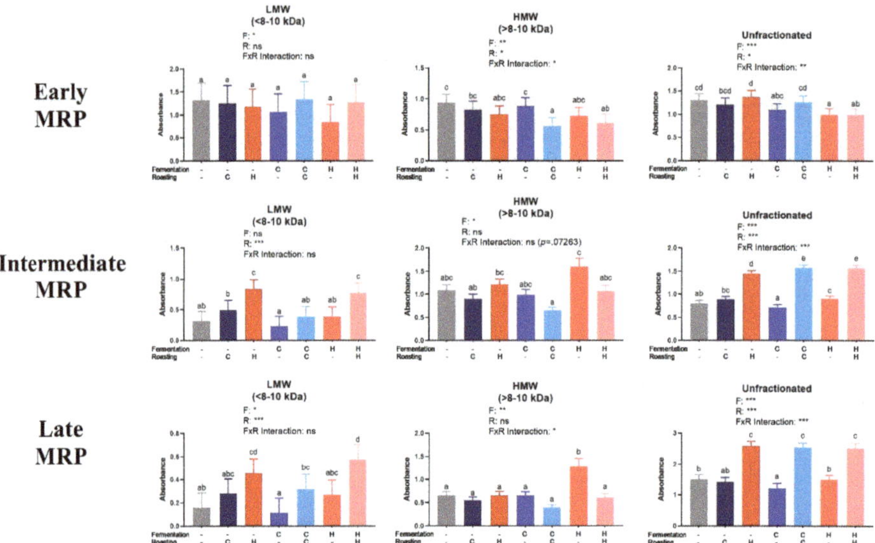

Figure 3. Analysis of low molecular weight (LMW) and high molecular weight (HMW) cocoa extract fractions and the unfractionated extract for early, intermediate, and late Maillard reaction products (MRP). Early MRP were quantified at 0.15625 mg/mL at 280 nm, intermediate MRP were quantified at 2.5 mg/mL at 360 nm, and late MRP were quantified at 5 mg/mL (LMW, <8–10 kDa), 2.5 mg/mL (HMW, >8–10 kDa), and 10 mg/mL (unfractionated extract) at 420 nm. Note that absolute absorbances are reported due to lack of adequate standards. Each bar represents the LS means with upper and lower CI. Significance between treatments was determined by two-way ANOVA for the roasting and fermentation temperature effects using type III sums of squares to account for unbalanced data, followed by post-hoc comparisons of LSMEANS ($p < 0.05$). Normality was checked for each variable visually and with the Shapiro–Wilks test, and, if needed, transformed (Log or Box-Cox) prior to running the ANOVA and post-hoc test. Treatments with different superscript letters (a–e) are significantly different within values. Legends above graphs indicate treatment (F= fermentation; R= roasting) main effect and interactions as determined by two-way ANOVA, * $p < 0.05$, ** $p < 0,01$, *** $p < 0.001$, refer to Table S4 for numerical values.

3.4. α-Glucosidase Enzyme Inhibition

Figure 4 shows the curves from four-parameter log-logistic model fitting of α-glucosidase activity as a function of inhibitor concentration for each treatment. See Figure S5 for raw (non-fitted) dose-response curves representing mean ± SEM values for each treatment. The extracts dose-dependently inhibited α-glucosidase activity. CF/CR was the most efficacious of all extracts and it was more potent than acarbose. At 250 µg/mL, all of the treatments but UF/HR and HF/HR inhibited enzyme activity greater than acarbose, and at concentrations ≥500 µg/mL, all the treatments were better inhibitors than acarbose (Figure 4). The extracts had the following IC$_{50}$ values from most potent to least potent: 68.09 (CF/CR) > 115.15 (HF/UR) > 125.31 (UF/CR) > 134.27 (CF/UR) > 154.14 (UF/UR) > 158.33 (HF/HR) > 169.19 (UF/HR), with acarbose (A) exhibiting an IC$_{50}$ value of 133.22 for comparison. Note that statistical comparisons of IC50 were not possible, since this is a parameter that is interpolated from a single curve for each treatment, fitted to multiple replicates at each dose. The curve parameters that were obtained for the four-parameter log-logistic model were also calculated for each treatment: Hill coefficient, minimum/maximum values, and EC_{50}, so that the statistical significance of inhibition parameters could be determined (Table 2). EC_{50} trends generally mimicked those of IC_{50} values.

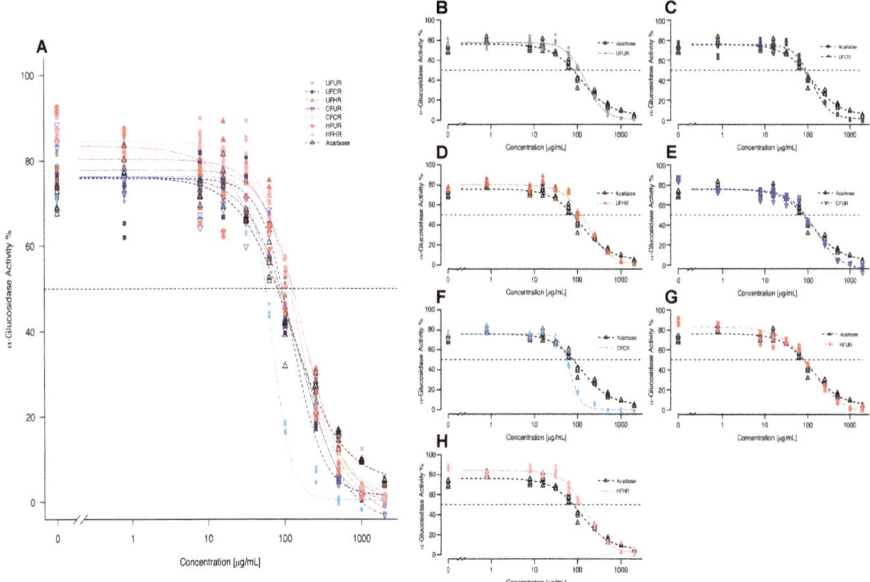

Figure 4. Overlay fitted curves obtained for α-glucosidase activity (% activity compared to no inhibitor) via 4-parameter log-logistic model for all cocoa powder extracts (**A**) and treatments plotted individually against acarbose (**B–H**). Individual points represent inhibition values for individual replicates at each concentration. Dotted line represents IC_{50} values. Refer to Table 2 for numerical values.

Table 2. Enzyme inhibition as analyzed by four-parameter log-logistic model. Significance indicated by * $p < 0.05$, ** $p < 0.01$, *** $p < 0.001$.

Parameter	Treatment			Estimate	Std. Error	LCL (2.5%)	UCL (97.5%)	*t*-Value	*p*-Value	Significance
	F	R	Abbreviation							
	-	-	(UF/UR)	1.9696	0.141725	1.691	2.248	13.8973	$<2.2 \times 10^{-16}$	***
	-	Cool	(UF/CR)	2.142171	0.177728	1.793	2.491	12.0531	$<2.2 \times 10^{-16}$	***
	-	Hot	(UF/HR)	1.603793	0.113242	1.381	1.826	14.1626	$<2.2 \times 10^{-16}$	***
Hill	Cool	-	(CF/UR)	1.429009	0.129739	1.174	1.684	11.0145	$<2.2 \times 10^{-16}$	***
Coefficient	Cool	Cool	(CF/CR)	3.465039	0.425681	2.629	4.301	8.14	3.17×10^{-16}	***
	Hot	-	(HF/UR)	1.142063	0.107241	0.931	1.353	10.6495	$<2.2 \times 10^{-16}$	***
	Hot	Hot	(HF/HR)	1.627053	0.113836	1.403	1.851	14.2929	$<2.2 \times 10^{-16}$	***
	Acarbose		(A)	1.30504	0.113057	1.083	1.527	11.5432	$<2.2 \times 10^{-16}$	***
	-	-	(UF/UR)	0.378196	1.496104	−2.561	3.318	0.2528	0.80054	
	-	Cool	(UF/CR)	1.513805	1.388439	−1.214	4.242	1.0903	0.27611	
	-	Hot	(UF/HR)	−0.030323	1.857523	−3.68	3.619	−0.0163	0.98698	
Minimum	Cool	-	(CF/UR)	−4.888001	2.078745	−8.972	−0.804	−2.3514	0.01909	*
Value	Cool	Cool	(CF/CR)	0.516432	0.977596	−1.404	2.437	0.5283	0.59755	
	Hot	-	(HF/UR)	−5.208691	2.488998	−10.1	−0.318	−2.0927	0.03689	*
	Hot	Hot	(HF/HR)	1.208122	1.755526	−2.241	4.657	0.6882	0.49166	
	Acarbose		(A)	4.867636	2.043044	0.854	8.882	2.3825	0.01757	*

Table 2. *Cont.*

Parameter	Treatment			Estimate	Std. Error	LCL (2.5%)	UCL (97.5%)	t-Value	p-Value	Significance
	F	R	Abbreviation							
Maximum Value	-	-	(UF/UR)	77.81598	0.894435	76.059	79.573	87.0002	$<2.2 \times 10^{-16}$	***
	-	Cool	(UF/CR)	75.880224	0.889772	74.132	77.628	85.2806	$<2.2 \times 10^{-16}$	***
	-	Hot	(UF/HR)	80.391462	0.940686	78.543	82.24	85.4604	$<2.2 \times 10^{-16}$	***
	Cool	-	(CF/UR)	76.297669	1.107626	74.121	78.474	68.8839	$<2.2 \times 10^{-16}$	***
	Cool	Cool	(CF/CR)	76.360965	0.968329	74.458	78.264	78.8585	$<2.2 \times 10^{-16}$	***
	Hot	-	(HF/UR)	83.518081	1.382998	80.801	86.235	60.3892	$<2.2 \times 10^{-16}$	***
	Hot	Hot	(HF/HR)	84.494321	0.96061	82.607	86.382	87.959	$<2.2 \times 10^{-16}$	***
	Acarbose		(A)	76.123584	1.073488	74.014	78.233	70.9123	$<2.2 \times 10^{-16}$	***
EC_{50}	-	-	(UF/UR)	153.38225	8.067667	137.53	169.23	19.012	$<2.2 \times 10^{-16}$	***
	-	Cool	(UF/CR)	122.95011	6.486545	110.21	135.7	18.9546	$<2.2 \times 10^{-16}$	***
	-	Hot	(UF/HR)	169.26537	10.618809	148.4	190.13	15.9401	$<2.2 \times 10^{-16}$	***
	Cool	-	(CF/UR)	146.08776	9.519332	127.38	164.79	15.3464	$<2.2 \times 10^{-16}$	***
	Cool	Cool	(CF/CR)	67.824539	1.888888	64.113	71.536	35.9071	$<2.2 \times 10^{-16}$	***
	Hot	-	(HF/UR)	127.62786	9.183673	109.58	145.67	13.8973	$<2.2 \times 10^{-16}$	***
	Hot	Hot	(HF/HR)	155.5339	8.935954	137.98	173.09	17.4054	$<2.2 \times 10^{-16}$	***
	Acarbose		(A)	119.95973	10.225587	99.869	140.05	11.7313	$<2.2 \times 10^{-16}$	***
IC_{50} [a]	-	-	(UF/UR)	154.1448	-	132.33 [a]	175.92 [b]	-	-	-
	-	Cool	(UF/CR)	125.30945	-	108.24 [a]	142.15 [b]	-	-	-
	-	Hot	(UF/HR)	169.1858	-	139.08 [a]	199.97 [b]	-	-	-
	Cool	-	(CF/UR)	134.26837	-	105.91 [a]	162.82 [b]	-	-	-
	Cool	Cool	(CF/CR)	68.09163	-	63.22 [a]	72.61 [b]	-	-	-
	Hot	-	(HF/UR)	115.14538	-	86.24 [a]	144.88 [b]	-	-	-
	Hot	Hot	(HF/HR)	158.33224	-	132.88 [a]	184.1 [b]	-	-	-
	Acarbose		(A)	133.22093	-	102.04 [a]	165.78 [b]	-	-	-

[a] IC_{50} LCL at 5%; [b] IC_{50} UCL at 95%.

3.5. Identifying Predictors of α-Glucosidase Enzyme Inhibitory Activity

Simple linear regression was employed to determine whether strong linear relationships existed between the observed α-glucosidase inhibitory activities (IC_{50}) and any of the individual chemical composition parameters of the cocoas and their extracts (Figure 5), as a means of tentatively identifying the composition factors that may influence inhibitory activities of cocoa. A strong negative correlation ($R = -0.882$) with a statistically significant non-zero slope was observed between the overall mDP calculated when factoring in monomers originally present in the sample, and IC_{50}. This demonstrates that the enzyme activity decreases as overall mDP increases (i.e. inhibition increases). No other composition value (total polyphenols, total flavanols, individual PCs, etc.) was strongly correlated to α-glucosidase inhibitory activity or had a non-zero slope. However, moderate positive correlations with IC_{50} were observed for the early and intermediate MRPs from the HMW extract fraction ($R \geq 0.52$ for both), demonstrating that the enzyme activity also increases as the concentration of these very large and complex compounds increase (i.e. inhibition decreases), which suggests that these MRPs either interfere with enzyme inhibition, or are markers of the loss of compounds that inhibit enzyme activity. Weak negative correlations with IC_{50} were observed for total flavanols, total PCs measured by HILIC, PC dimers, tetramers, and nonamers and early MRPs (LMW fraction) ($|R| > 0.2$ for all). Finally, weak positive correlations with IC_{50} were observed for intermediate MRPs (LMW) and late MRPs (HMW).

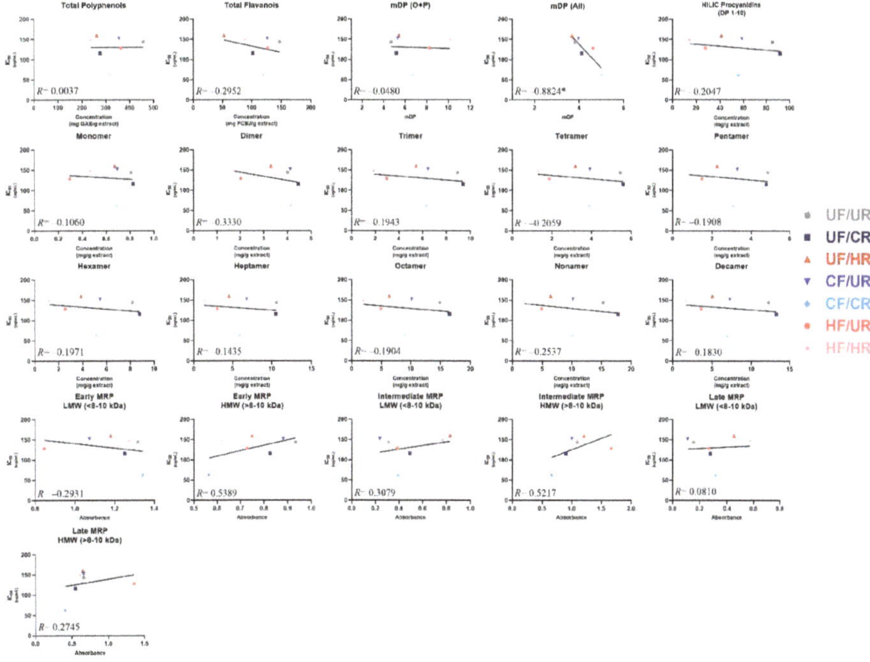

Figure 5. Correlations between cocoa powder extract composition and enzyme IC$_{50}$ values. For mean degree of polymerization (mDP), O + P: oligomers + polymers (not factoring in monomers present prior to thiolysis), and All: (including monomers present prior to thiolysis). Note the x-axis for mDP graphs have a minimum value of 1 mDP because there cannot be an mDP value of <1. Note that early, intermediate, and late MRP are presented as absolute absorbance. Individual points represent mean composition values and calculated IC$_{50}$ values for each treatment. Lines represent the least-squares regression line for each plot. Note that composition values are for the cocoa extract, not powders, since the extract was evaluated for enzyme activity. *indicates that the slope is significantly different from zero.

4. Discussion

We generated seven different cocoa powders representing a broad range of possible fermentation and roasting conditions, and then assessed compositional and bioactivity differences of these powders while using a partial factorial design through controlled fermentation and further processing. This novel approach allowed for us to solely attribute differences to processing, since all of the products were produced from the same starting beans.

Cocoa fermentation, as mentioned previously, is reliant upon multiple highly variable factors, and thus fermentation conditions employed in cocoa production around the world vary tremendously. The reported heap fermentations appear to start uniformly around pH 3.6, but our model system started slightly out of these ranges [27–29,38–40]. Our cool and hot fermentations both concluded under more acidic conditions than is typically reported. Beans of Trinitario and Forastero varieties typically range in initial pH from 6.3–6.8, decreasing to approximately 5.0–6.0 by the end of fermentation, whereas Criollo beans, while not as widely produced or studied as the Trinitario and Forastero varieties, have been noted for their characteristically low pH [39–41]. The high initial DO values can be attributed to the fresh mixing of bean and simulated pulp media, dropping significantly after 24 h and remaining ≤1 mg/L for the remainder of the fermentation. This model system fermentation is liquid based and it was designed to mimic typical conditions in the center of a well-mixed cocoa heap. The FI and cut test are traditional quality controls conducted on farm to assess the post-fermentation bean quality. These methods are based upon anthocyanin degradation and aglycone release throughout fermentation. The FI of raw unfermented beans typically range from 0.3–0.6 and increase to 1.3–1.4 during fermentation, with a value of ≥1 indicating the near-complete anthocyanin degradation and, thus, adequate and complete fermentation [2,42,43]. However, our FI values started at >1 (Figure 1B). These data, along with the acidic values for both pulp and bean in CF and HF systems, lead us to believe that these beans are of Criollo or Nacional variety, but this statement cannot be verified and is thus a limitation of this study. Criollo beans have low levels of anthocyanins when compared to other varieties, like Trinitario and Forastero, and they have a naturally low pH, which possibly explains the high acidity in our fermentation systems and inconclusive FI and cut test results [41]. Nacional beans grow in Ecuador and they are very similar to Criollo beans [41]. Traceability is often limited or not possible in a global commodity supply chain, such as that of cocoa beans, and thus the exact variety of the beans used in this study is unknown.

The total polyphenol and flavanol content differed between powder and bean products, with powders having higher total concentrations of both polyphenols and flavanols due to concentration of flavanols in powders via the removal of flavanol-free cocoa butter during pressing. Overall, our cocoa composition data align with previously published reports that fermentation and roasting significantly reduce native flavanol levels in cocoa beans [5,11,14,17,32,44–46]. We assessed the interaction of both steps by evaluating them in varying combinations, while most previous studies have investigated the implications of fermentation and roasting independently of each other. Roasting is often considered to be the key phase in cocoa processing in terms of defining the sensory characteristics of a finished product by producing characteristic aromas, flavors, and texture of beans. Yet, roasting is an extension of the flavanol reduction that begins during fermentation. The epimerization and polymerization of flavanols, as well as reactions with larger structures, such as proteins, polysaccharides, and MRPs, can all influence the flavonoid levels in processed cocoa [5,46]. This is clearly demonstrated by our data across a broad range of polyphenol and flavanol assays.

Significantly lower levels of all measured compounds in the most harshly processed cocoa, HF/HR, as compared to other treatments, demonstrate that prolonged high temperature exposure via hot fermentation, followed by hot roasting degrades native polyphenols and flavanols. However, the high mDP that was reported for both the HF/UR and HF/HR suggests that fermentation might have a larger influence on polymerization of larger molecular weight compounds than originally thought, or that the interaction between fermentation and roasting at high temperatures is crucial in the development of large PCs. This agrees with studies suggesting that levels of high molecular weight PCs increase with greater roasting time and temperatures [10,17].

Flavanols, particularly HMW flavanols, generally have poor intestinal absorption, and we chose to examine the inhibition of a digestive enzyme, such as α-glucosidase (a brush-border enzyme), in order to examine a mechanism located where these compounds are present at their highest concentrations in vivo (the lumen or epithelium of the intestines). We also chose α-glucosidase based on data suggesting that cocoa exerts greater inhibition on this enzyme than on the pancreatic α-amylase and lipase [13]. Through the inhibition of α-glucosidase, cocoa compounds have the potential to slow down carbohydrate digestion and post-prandial absorption of glucose, which blunts blood glucose excursions. Extracts that were produced in this study appear to be effective dietary inhibitors of α-glucosidase, most notably the CF/CR treatment, with an IC_{50} of 68.1 μg/mL (~50% lower than that of acarbose). The level of acarbose that is typically present in the gut is between 25–100 μg/mL when taken as recommended, which is within the concentrations used in the present study, and thus represents a relevant control for this assay. At an IC_{50} of 68.1 μg/mL, CF/CR falls within the typical acarbose range, but it is two-fold more effective in inhibiting 50% of enzymatic activity. The CF/CR processing parameters that were applied in this study were within ranges used in industrial cocoa powder production and have promising potential to surpass the activity of acarbose in vivo. In addition to CF/CR, HF/UR, and UF/CR also had lower IC_{50} values than acarbose. Furthermore, all of the cocoas were more effective than acarbose at higher doses, such as 500 µg/mL (Figure 4). While 500 µg/mL is approximately five-fold higher than typical acarbose concentrations within the gut, it is still a highly relevant dietary dose at 6.67 g of original cocoa product or approximately 0.24 square of baking chocolate. Even at 250 µg/mL, all the cocoas treatments, except UF/HR and HF/HR, had better inhibitory effects than acarbose, and these were treatments that were not significantly different than acarbose (Figure 4). While it is promising that select processing conditions improve IC_{50} values, it is noteworthy that even the harshest of conditions have similar inhibitory activity as compared to UF/UR (no fermentation or roasting processing). The least-processed cocoa (UF/UR) was not the best inhibitor nor was the most harshly-processed cocoa (HF/HR) far worse than all other treatments, which suggests that, in this particular instance, processing, at worst, does not negatively affect activity and, at best, can actually greatly enhance inhibitory activity.

The Hill coefficient is the slope of the curve at the inflection point and it can be used to interpret the binding behavior and kinetic activity of the inhibitor's target. Furthermore, a Hill coefficient of 1 is considered to be standard and is generally understood to have a single inhibitor binding site or a simple kinetic mechanism (Table 2). With a steeper curve, the Hill coefficient increases, which indicated that enzyme inhibition increases with an overall decrease in concentration range. Although the reasons for these steep dose-response curves are poorly understood, several mechanisms can potentially explain this action, including an increased number of inhibitor sites, an inhibitor undergoing a phase transition with increased concentration, as well as when the enzyme concentration exceeds the equilibrium constant for the inhibitor. With the exception of UF/CR and CF/CR, all of the treatments have standard slopes and, therefore, can be considered to have simple kinetic mechanisms. EC_{50}, or the half-maximal effective concentration in relation to the control (i.e. acarbose), can be used in combination with log-logistic parameters to estimate the IC_{50} of each treatment. There are slight differences between these and IC_{50} values due to dependence on the control for EC_{50} values. Yet, CF/CR is the most powerful inhibitor of α-glucosidase with both EC_{50} and IC_{50} values being lower than all other treatments, including acarbose. Overall, our data do not support the hypothesis that an inverse relationship exists between the processing intensity and α-glucosidase inhibitory activity of cocoa powder.

Once we determined the α-glucosidase inhibitory activities of these cocoas, we then wished to determine whether measured concentrations of putative bioactive cocoa compounds were associated with inhibitory activity. We employed a simple linear regression approach to achieve this, similar to our previous studies [13,47] (Figure 5). Of all compositional measures, the only strong correlation seen was between IC_{50} and the overall flavanol mDP, with decreasing IC_{50} as mDP increased. The lack of strong correlations for most measures in Figure 5 aligns with previous work demonstrating that, although processing induced significant losses in total polyphenols and total flavanols, these compositional changes did not uniformly influence bioactivity, but, rather, increasing mDP had a stronger influence on cocoa bioactivity [13,25,48,49]. This finding provides evidence suggesting that cocoa processing could be specifically tailored to promote flavanol polymerization as a means to enhance α-glucosidase activity. Additionally, our data begin to suggest that MRPs may be useful as markers of activity in cocoa. These compounds are intriguing markers, as they may directly contribute to α-glucosidase inhibition due to structural similarities with their carbohydrate precursors, and they are also sensitive indicators of processes, such as roasting. The quantification of these compounds is complex, with limited understanding of structure and activity. Our results preliminarily suggest that longer and higher roasting times/temperatures, which often result in the increased production of these MRPs, negatively impact the α-glucosidase inhibitory activity of the extract, but further investigation into these compounds is needed to fully elucidate the impact that they have on specific digestive enzymes. Finally, our data suggest that traditional putative markers, such as total polyphenols and total or specific flavanols, may not be sufficient for predicting α-glucosidase inhibitory activity of cocoa. Whether this applies to other bioactivities, in vitro and in vivo, remains to be seen.

The main novelties of this study were in (1) our use of a single uniform batch of raw cocoa beans as the starting material for all the processing treatments, (2) coupled with examining variations in fermentation and roasting in combination as opposed to separately, and (3) challenging the hypothesis that traditional putative bioactives in cocoa are correlated with a given bioactivity. Previous studies have often relied on information that was reported through various levels of the cocoa supply chain, often resulting in unknowns regarding origin of the beans, processing conditions, as most cocoa production processes in country of origin, especially fermentation, lack robust controls or recordkeeping, which results in considerable variation and poor traceability. Our controlled system eliminates many of these external challenges, providing confidence that differences between the cocoa powders were due to the fermentation and roasting treatments, rather than unknown factors. While the model fermentation system that was used in our study was not designed to physically mimic the conditions found in on-farm cocoa fermentation, fermentation of cocoa beans while using this system results in biochemical changes analogous to those reported in on farm fermentation. Conducting the fermentation step in this model system allowed for us to produce cocoa powders with acceptable chemical composition, which were subject to known and controlled fermentation conditions. We are currently expanding on this research by investigating each cocoa powder's ability to prevent obesity-induced GI and systemic inflammation and gut barrier dysfunction in vivo.

This study is not without limitations. Quantifying individual PCs ranging from DP 1–10, although expanding beyond the monomeric flavan-3-ols often focused on in cocoa products, still leaves many large molecular weight compounds yet to be individually quantified. Additionally, α-glucosidase inhibition is just one specific mode of bioactivity that we chose to focus on, based on previous reports. Although our results begin to expand on this powerful inhibitory effect, the exact compositional factors that are involved remain to be elucidated. Our processing parameters could be expanded to include a wider range of processing conditions to address these limitations moving forward, to further optimize the α-glucosidase inhibitory activity. Additional work is also needed to study the finer variations of the optimal parameters that we identified here to further enhance the inhibitory activity. This expansion of treatment conditions tested, as well as further the fractionation of powder extracts to identify specific components that are associated with α-glucosidase inhibitory activity, would allow for a more comprehensive understanding of the impact that each processing step has on various compositional factors and bioactivity of dietary cocoa. By extending characterization to lignins and the speciation of melanoidins and other MRPs, further evidence could be provided to explain the mechanisms that govern enzymatic α-glucosidase inhibition by cocoa. This in vitro work also needs to be extended in vivo to further reinforce and clarify the mechanisms behind cocoa's enzymatic inhibition influence in both animals and humans, including, but not limited to, maltose versus glucose tolerance tests with cocoa consumption, and long-term effects on diet-induced obesity and glucose intolerance. Furthermore, yeast-derived α-glucosidase is a good, inexpensive starting point to screen for α-glucosidase inhibitory activity *in vitro*. Although yeast derived α-glucosidase is not completely identical to mammalian α-glucosidase, it is commonly selected for anti-diabetic investigations and it can be used for preliminary screening before investigating further in costly mammalian (i.e. rat acetone intestinal powder or Caco-2 cells) or human recombinant enzymes. For our purposes, yeast α-glucosidase was initially used to demonstrate the concept that processing does not necessarily eliminate cocoa's α-glucosidase inhibitory activity and, therefore, was not intended to be a definitive test of in vivo human relevance. Cocoa, as well as other polyphenol-rich substances, have exhibited powerful inhibitory activity when using mammalian α-glucosidase, showing promise for our data moving forward [16,50–52]. However, further studies must be conducted to establish relevance between these models and the human digestive tract, as well as specific mechanisms of action.

5. Conclusions

Overall, this study demonstrates that processed cocoa powders are promising inhibitors of α-glucosidase, despite a significant reduction in native flavanol composition during fermentation and roasting, and fermentation and roasting can improve inhibitory activity when compared to raw cocoa. We report the novel finding that cocoa processing might generate compounds with α-glucosidase inhibitory activity, and that non-traditional markers, such as MRPs, may be more informative than traditional markers, such as total and individual polyphenols and flavanols. These observations support our hypothesis that reductions in native polyphenols and flavanols do not necessarily dictate a reduction in activity and, furthermore, that products of fermentation and roasting do, in fact, contribute to cocoa bioactivity. Further investigation is needed to determine the identity of compounds that explain this activity. Finally, it remains to be seen whether these findings apply to other bioactivities of cocoa, but the present study provides the proof of concept needed to justify such investigations.

Supplementary Materials: The following are available online at http://www.mdpi.com/2076-3921/8/12/635/s1, Table S1: MS/MS settings for MRM detection of monomer-decamer flavanols, Table S2: Preliminary melanoidin identification, Table S3: Specifications for cocoa liquors and cakes of each treatment, Table S4: Compositional data analysis as determined by two-way ANOVA for roasting and fermentation effect using type III sums of squares to account for unbalanced data. Figure S1: Progression of one cool fermentation batch from 0–168 h, followed by bean oven drying, Figure S2: (A) Total polyphenols, expressed in gallic acid equivalents, and (B) total flavanols, expressed in procyanidin B2 equivalents, of cocoa beans. Figure S3: Levels of individual procyanidin compounds in cocoa beans, as quantified by HILIC UPLC-MS/MS. Figure S4: Levels of individual procyanidin compounds in cocoa powders, as quantified by HILIC UPLC-MS/MS. Figure S5: Dose response curve for α-glucosidase activity (% activity compared to no inhibitor) for cocoa powder extracts.

Author Contributions: Conceptualization, J.D.L., A.C.S., A.P.N.; Methodology, J.D.L., A.C.S., A.P.N., B.D.W., K.C.R.; Formal Analysis, A.P.N., K.C.R., A.C.S., A.H.L.; Data Curation, K.C.R., L.E.G., L.A.E., H.H.; Visualization, H.H.; Writing—Original Draft Preparation, K.C.R., A.P.N.; Writing—Review & Editing, K.C.R., A.P.N., J.D.L., A.C.S.

Funding: This work was supported by USDA AFRI Grant 2017-67017-26783. Funding for this work was also provided, in part, by the Virginia Agricultural Experiment Station and the Hatch Program of the National Institute of Food and Agriculture, U.S. Department of Agriculture.

Acknowledgments: The authors would like to acknowledge Wyatt Elder at Cargill Cocoa for his generous donation of raw unfermented cocoa beans, Sebastian Wolfrum, James Naquin, and Lea Woodard at Epiphany Craft Malt in Durham, NC for assistance in roasting, and Arlen Moser, Matt Mastrog, and the Blommer Chocolate Research and Development team in East Greenville, PA for their time and guidance in processing all cocoa powders, free of charge. Additionally, the authors would like to thank Hengjian Wang for his help in freeze drying various samples and Amy Moore for help in fermentation sample preparation.

Conflicts of Interest: The authors declare no conflict of interest.

References

1. Andújar, I.; Recio, M.C.; Giner, R.M.; Ríos, J.L. Cocoa polyphenols and their potential benefits for human health. *Oxid. Med. Cell. Longev.* **2012**, *2012*. [CrossRef] [PubMed]
2. Kongor, J.E.; Hinneh, M.; de Walle, D.V.; Afoakwa, E.O.; Boeckx, P.; Dewettinck, K. Factors influencing quality variation in cocoa (*Theobroma cacao*) bean flavour profile—A review. *Food Res. Int.* **2016**, *82*, 44–52. [CrossRef]
3. Schinella, G.; Mosca, S.; Cienfuegos-Jovellanos, E.; Pasamar, M.A.; Muguerza, B.; Ramon, D.; Rios, J.L. Antioxidant properties of polyphenol-rich cocoa products industrially processed. *Food Res. Int.* **2010**, *43*, 1614–1623. [CrossRef]
4. Hatano, T.; Miyatake, H.; Natsume, M.; Osakabe, N.; Takizawa, T.; Ito, H.; Yoshida, T. Proanthocyanidin glycosides and related polyphenols from cacao liquor and their antioxidant effects. *Phytochemistry* **2002**, *59*, 749–758. [CrossRef]
5. Oracz, J.; Nebesny, E.; Żyżelewicz, D. Changes in the flavan-3-ols, anthocyanins, and flavanols composition of cocoa beans of different *Theobroma cacao* L. groups affected by roasting conditions. *Eur. Food Res. Technol.* **2015**, *241*, 663–681. [CrossRef]
6. Cooper, K.A.; Donovan, J.L.; Waterhouse, A.L.; Williamson, G. Cocoa and health: A decade of research. *Br. J. Nutr.* **2008**, *99*, 1–11. [CrossRef]
7. Wollgast, J.; Anklam, E. Review on polyphenols in *Theobroma cacao*: Changes in composition during the manufacture of chocolate and methodology for identification and quantification. *Food Res. Int.* **2000**, *33*, 423–447. [CrossRef]

8. de Brito, E.S.; García, N.H.P.; Gallão, M.I.; Cortelazzo, A.L.; Fevereiro, P.S.; Braga, M.R. Structural and chemical changes in cocoa (*Theobroma cacao* L) during fermentation, drying and roasting. *J. Sci. Food Agric.* **2001**, *81*, 281–288. [CrossRef]

9. Mazor Jolić, S.; Radojčić Redovniković, I.; Marković, K.; Ivanec Šipušić, Đ.; Delonga, K. Changes of phenolic compounds and antioxidant capacity in cocoa beans processing. *Int. J. Food Sci. Technol.* **2011**, *46*, 1793–1800. [CrossRef]

10. Oracz, J.; Nebesny, E. Effect of roasting parameters on the physicochemical characteristics of high-molecular-weight Maillard reaction products isolated from cocoa beans of different *Theobroma cacao* L. groups. *Eur. Food Res. Technol.* **2019**, *245*, 111–128. [CrossRef]

11. Albertini, B.; Schoubben, A.; Guarnaccia, D.; Pinelli, F.; Della Vecchia, M.; Ricci, M.; Di Renzo, G.C.; Blasi, P. Effect of fermentation and drying on cocoa polyphenols. *J. Agric. Food Chem.* **2015**, *63*, 9948–9953. [CrossRef] [PubMed]

12. Dorenkott, M.R.; Griffin, L.E.; Goodrich, K.M.; Thompson-Witrick, K.A.; Fundaro, G.; Ye, L.; Stevens, J.R.; Ali, M.; O'Keefe, S.F.; Hulver, M.W.; et al. Oligomeric Cocoa Procyanidins Possess Enhanced Bioactivity Compared to Monomeric and Polymeric Cocoa Procyanidins for Preventing the Development of Obesity, Insulin Resistance, and Impaired Glucose Tolerance during High-Fat Feeding. *J. Agric. Food Chem.* **2014**, *62*, 2216–2227. [CrossRef] [PubMed]

13. Ryan, C.M.; Khoo, W.; Ye, L.; Lambert, J.D.; O'Keefe, S.F.; Neilson, A.P. Loss of Native Flavanols during Fermentation and Roasting Does Not Necessarily Reduce Digestive Enzyme-Inhibiting Bioactivities of Cocoa. *J. Agric. Food Chem.* **2016**, *64*, 3616–3625. [CrossRef] [PubMed]

14. Di Mattia, C.; Martuscelli, M.; Sacchetti, G.; Scheirlinck, I.; Beheydt, B.; Mastrocola, D.; Pittia, P. Effect of Fermentation and Drying on Procyanidins, Antiradical Activity and Reducing Properties of Cocoa Beans. *Food Bioprocess Technol.* **2013**, *6*, 3420–3432. [CrossRef]

15. Quiroz-Reyes, C.N.; Fogliano, V. Design cocoa processing towards healthy cocoa products: The role of phenolics and melanoidins. *J. Funct. Foods* **2018**, *45*, 480–490. [CrossRef]

16. Bellesia, A.; Tagliazucchi, D. Cocoa brew inhibits in vitro α-glucosidase activity: The role of polyphenols and high molecular weight compounds. *Food Res. Int.* **2014**, *63*, 439–445. [CrossRef]

17. Ioannone, F.; Di Mattia, C.D.; De Gregorio, M.; Sergi, M.; Serafini, M.; Sacchetti, G. Flavanols, proanthocyanidins and antioxidant activity changes during cocoa (*Theobroma cacao* L.) roasting as affected by temperature and time of processing. *Food Chem.* **2015**, *174*, 256–262. [CrossRef]

18. Summa, C.; McCourt, J.; Cämmerer, B.; Fiala, A.; Probst, M.; Kun, S.; Anklam, E.; Wagner, K.-H. Radical scavenging activity, anti-bacterial and mutagenic effects of Cocoa bean Maillard Reaction products with degree of roasting. *Mol. Nutr. Food Res.* **2008**, *52*, 342–351. [CrossRef]

19. Summa, C.; Raposo, F.C.; McCourt, J.; Scalzo, R.L.; Wagner, K.-H.; Elmadfa, I.; Anklam, E. Effect of roasting on the radical scavenging activity of cocoa beans. *Eur. Food Res. Technol.* **2006**, *222*, 368–375. [CrossRef]

20. Donovan, J.L.; Lee, A.; Manach, C.; Rios, L.; Morand, C.; Scalbert, A.; Rémésy, C. Procyanidins are not bioavailable in rats fed a single meal containing a grapeseed extract or the procyanidin dimer B 3. *Br. J. Nutr.* **2002**, *87*, 299–306. [CrossRef]

21. Serra, A.; Macia, A.; Romero, M.-P.; Valls, J.; Bladé, C.; Arola, L.; Motilva, M.-J. Bioavailability of procyanidin dimers and trimers and matrix food effects in in vitro and in vivo models. *Br. J. Nutr.* **2010**, *103*, 944–952. [CrossRef] [PubMed]

22. Ottaviani, J.I.; Kwik-Uribe, C.; Keen, C.L.; Schroeter, H. Intake of dietary procyanidins does not contribute to the pool of circulating flavanols in humans. *Am. J. Clin. Nutr.* **2012**, *95*, 851–858. [CrossRef] [PubMed]

23. Crozier, A.; Del Rio, D.; Clifford, M.N. Bioavailability of dietary flavonoids and phenolic compounds. *Mol. Aspects Med.* **2010**, *31*, 446–467. [CrossRef] [PubMed]

24. Wiese, S.; Esatbeyoglu, T.; Winterhalter, P.; Kruse, H.-P.; Winkler, S.; Bub, A.; Kulling, S.E. Comparative biokinetics and metabolism of pure monomeric, dimeric, and polymeric flavan-3-ols: A randomized cross-over study in humans. *Mol. Nutr. Food Res.* **2015**, *59*, 610–621. [CrossRef] [PubMed]

25. Gu, Y.; Hurst, W.J.; Stuart, D.A.; Lambert, J.D. Inhibition of Key Digestive Enzymes by Cocoa Extracts and Procyanidins. *J. Agric. Food Chem.* **2011**, *59*, 5305–5311. [CrossRef]

26. Zhang, L.-L.; Han, L.; Yang, S.-Y.; Meng, X.-M.; Ma, W.-F.; Wang, M. The mechanism of interactions between flavan-3-ols against a-glucosidase and their in vivo antihyperglycemic effects. *Bioorganic Chem.* **2019**, *85*, 364–372. [CrossRef]

27. Camu, N.; Winter, T.D.; Verbrugghe, K.; Cleenwerck, I.; Vandamme, P.; Takrama, J.S.; Vancanneyt, M.; Vuyst, L.D. Dynamics and Biodiversity of Populations of Lactic Acid Bacteria and Acetic Acid Bacteria Involved in Spontaneous Heap Fermentation of Cocoa Beans in Ghana. *Appl. Environ. Microbiol.* **2007**, *73*, 1809–1824. [CrossRef]

28. Camu, N.; González, A.; De Winter, T.; Van Schoor, A.; De Bruyne, K.; Vandamme, P.; Takrama, J.S.; Addo, S.K.; De Vuyst, L. Influence of turning and environmental contamination on the dynamics of populations of lactic acid and acetic acid bacteria involved in spontaneous cocoa bean heap fermentation in Ghana. *Appl. Environ. Microbiol.* **2008**, *74*, 86–98. [CrossRef]

29. Schwan, R.F.; Wheals, A.E. The Microbiology of Cocoa Fermentation and its Role in Chocolate Quality. *Crit. Rev. Food Sci. Nutr.* **2004**, *44*, 205–221. [CrossRef]

30. Racine, K.C.; Lee, A.H.; Wiersema, B.D.; Huang, H.; Lambert, J.D.; Stewart, A.C.; Neilson, A.P. Development and Characterization of a Pilot-Scale Model Cocoa Fermentation System Suitable for Studying the Impact of Fermentation on Putative Bioactive Compounds and Bioactivity of Cocoa. *Foods* **2019**, *8*, 102. [CrossRef]

31. Lee, A.H.; Neilson, A.P.; O'Keefe, S.F.; Ogejo, J.A.; Huang, H.; Ponder, M.; Chu, H.S.S.; Jin, Q.; Pilot, G.; Stewart, A.C. A laboratory-scale model cocoa fermentation using dried, unfermented beans and artificial pulp can simulate the microbial and chemical changes of on-farm cocoa fermentation. *Eur. Food Res. Technol.* **2018**. [CrossRef]

32. Camu, N.; De Winter, T.; Addo, S.K.; Takrama, J.S.; Bernaert, H.; De Vuyst, L. Fermentation of cocoa beans: influence of microbial activities and polyphenol concentrations on the flavour of chocolate. *J. Sci. Food Agric.* **2008**, *88*, 2288–2297. [CrossRef]

33. Schwan, R.F. Cocoa fermentations conducted with a defined microbial cocktail inoculum. *Appl. Environ. Microbiol.* **1998**, *64*, 1477–1483. [PubMed]

34. Romero-Cortes, T.; Salgado-Cervantes, M.A.; García-Alamilla, P.; García-Alvarado, M.A.; del C Rodríguez-Jimenes, G.; Hidalgo-Morales, M.; Robles-Olvera, V. Relationship between fermentation index and other biochemical changes evaluated during the fermentation of Mexican cocoa (*Theobroma cacao*) beans. *J. Sci. Food Agric.* **2013**, *93*, 2596–2604. [CrossRef]

35. Racine, K.C.; Lee, A.H.; Stewart, A.C.; Blakeslee, K.W.; Neilson, A.P. Development of a rapid ultra performance hydrophilic interaction liquid chromatography tandem mass spectrometry method for procyanidins with enhanced ionization efficiency. *J. Chromatogr. A* **2019**, *1594*, 54–64. [CrossRef]

36. Sacchetti, G.; Ioannone, F.; De Gregorio, M.; Di Mattia, C.; Serafini, M.; Mastrocola, D. Non enzymatic browning during cocoa roasting as affected by processing time and temperature. *J. Food Eng.* **2016**, *169*, 44–52. [CrossRef]

37. Balfour, J.A.; McTavish, D. Acarbose. *Drugs* **1993**, *46*, 1025–1054. [CrossRef]

38. Nielsen, D.S.; Teniola, O.D.; Ban-Koffi, L.; Owusu, M.; Andersson, T.S.; Holzapfel, W.H. The microbiology of Ghanaian cocoa fermentations analysed using culture-dependent and culture-independent methods. *Int. J. Food Microbiol.* **2007**, *114*, 168–186. [CrossRef]

39. Ho, V.T.T.; Zhao, J.; Fleet, G. The effect of lactic acid bacteria on cocoa bean fermentation. *Int. J. Food Microbiol.* **2015**, *205*, 54–67. [CrossRef]

40. Ardhana, M.M.; Fleet, G.H. The microbial ecology of cocoa bean fermentations in Indonesia. *Int. J. Food Microbiol.* **2003**, *86*, 87–99. [CrossRef]

41. Saltini, R.; Akkerman, R.; Frosch, S. Optimizing chocolate production through traceability: A review of the influence of farming practices on cocoa bean quality. *Food Control* **2013**, *29*, 167–187. [CrossRef]

42. De Taeye, C.; Eyamo Evina, V.J.; Caullet, G.; Niemenak, N.; Collin, S. Fate of anthocyanins through cocoa fermentation. Emergence of new polyphenolic dimers. *J. Agric. Food Chem.* **2016**, *64*, 8876–8885. [CrossRef] [PubMed]

43. Emmanuel, O.A.; Jennifer, Q.; Agnes, S.B.; Jemmy, S.T.; Firibu, K.S. Influence of pulp-preconditioning and fermentation on fermentative quality and appearance of Ghanaian cocoa (*Theobroma cacao*) beans. *Int. Food Res. J. Selangor.* **2012**, *19*, 127–133.

44. Payne, M.J.; Hurst, W.J.; Miller, K.B.; Rank, C.; Stuart, D.A. Impact of Fermentation, Drying, Roasting, and Dutch Processing on Epicatechin and Catechin Content of Cacao Beans and Cocoa Ingredients. *J. Agric. Food Chem.* **2010**, *58*, 10518–10527. [CrossRef]

45. Kim, H.; Keeney, P.G. (−)-Epicatechin Content in Fermented and Unfermented Cocoa Beans. *J. Food Sci.* **1984**, *49*, 1090–1092. [CrossRef]

46. Kothe, L.; Zimmermann, B.F.; Galensa, R. Temperature influences epimerization and composition of flavanol monomers, dimers and trimers during cocoa bean roasting. *Food Chem.* **2013**, *141*, 3656–3663. [CrossRef]

47. Ryan, C.M.; Khoo, W.; Stewart, A.C.; O'Keefe, S.F.; Lambert, J.D.; Neilson, A.P. Flavanol concentrations do not predict dipeptidyl peptidase-IV inhibitory activities of four cocoas with different processing histories. *Food Funct.* **2017**, *8*, 746–756. [CrossRef]

48. Zhong, H.; Xue, Y.; Lu, X.; Shao, Q.; Cao, Y.; Wu, Z.; Chen, G. The Effects of Different Degrees of Procyanidin Polymerization on the Nutrient Absorption and Digestive Enzyme Activity in Mice. *Molecules* **2018**, *23*, 2916. [CrossRef]

49. Yamashita, Y.; Okabe, M.; Natsume, M.; Ashida, H. Comparison of anti-hyperglycemic activities between low-and high-degree of polymerization procyanidin fractions from cacao liquor extract. *J. Food Drug Anal.* **2012**, *20*, 283–287.

50. Hogan, S.; Zhang, L.; Li, J.; Sun, S.; Canning, C.; Zhou, K. Antioxidant rich grape pomace extract suppresses postprandial hyperglycemia in diabetic mice by specifically inhibiting alpha-glucosidase. *Nutr. Metab.* **2010**, *7*, 71. [CrossRef]

51. Lavelli, V.; Harsha, P.S.; Ferranti, P.; Scarafoni, A.; Iametti, S. Grape skin phenolics as inhibitors of mammalian α-glucosidase and α-amylase–effect of food matrix and processing on efficacy. *Food Funct.* **2016**, *7*, 1655–1663. [CrossRef] [PubMed]

52. McDougall, G.J.; Shpiro, F.; Dobson, P.; Smith, P.; Blake, A.; Stewart, D. Different polyphenolic components of soft fruits inhibit α-amylase and α-glucosidase. *J. Agric. Food Chem.* **2005**, *53*, 2760–2766. [CrossRef] [PubMed]

 *antioxidants*

Article

The Kinetics of Total Phenolic Content and Monomeric Flavan-3-ols during the Roasting Process of Criollo Cocoa

Editha Fernández-Romero [1], Segundo G. Chavez-Quintana [2], Raúl Siche [3], Efraín M. Castro-Alayo [2,4,*] and Fiorella P. Cardenas-Toro [4]

[1] Programa Académico de Ingeniería Agroindustrial, Facultad de Ingeniería y Ciencias Agrarias, Universidad Nacional Toribio Rodríguez de Mendoza de Amazonas, Calle Higos Urco 342-350-356, Chachapoyas, Amazonas, Peru; fer.virgo59@gmail.com

[2] Instituto de Investigación, Innovación y Desarrollo para el Sector Agrario y Agroindustrial de la Región Amazonas (IIDAA-Amazonas), Facultad de Ingeniería y Ciencias Agrarias, Universidad Nacional Toribio Rodríguez de Mendoza de Amazonas, Calle Higos Urco 342-350-356, Chachapoyas, Amazonas, Peru; segundo.quintana@untrm.edu.pe

[3] Facultad de Ciencias Agropecuarias, Universidad Nacional de Trujillo, Av. Juan Pablo II s/n, Ciudad Universitaria, Trujillo 13001, Peru; rsiche@unitru.edu.pe

[4] Sección de Ingeniería Industrial, Departamento de Ingeniería, Pontificia Universidad Católica del Perú, Av. Universitaria 1801, San Miguel 150136, Lima 32, Peru; fcardenas@pucp.pe

* Correspondence: efrain.castro@untrm.edu.pe or efrain.castro@pucp.edu.pe; Tel.: +51-986-376-463

Received: 27 November 2019; Accepted: 15 January 2020; Published: 9 February 2020

Abstract: Cocoa beans are the main raw material for the manufacture of chocolate and are currently gaining great importance due to their antioxidant potential attributed to the total phenolic content (TPC) and the monomeric flavan-3-ols (epicatechin and catechin). The objective of this study was to determine the degradation kinetics parameters of TPC, epicatechin, and catechin during the roasting process of Criollo cocoa for 10, 20, 30, 40, and 50 min at 90, 110, 130, 150, 170, 190, and 200 °C. The results showed a lower degradation of TPC (10.98 ± 6.04%) and epicatechin (8.05 ± 3.01%) at 130 °C and 10 min of roasting, while a total degradation of epicatechin and a 92.29 ± 0.06% degradation of TPC was obtained at 200 °C and 50 min. Reaction rate constant (k) and activation energy (E_a) were 0.02–0.10 min^{-1} and 24.03 J/mol for TPC and 0.02–0.13 min^{-1} and 22.51 J/mol for epicatechin, respectively. Degradation kinetics of TPC and epicatechin showed first-order reactions, while the catechin showed patterns of formation and degradation.

Keywords: roasting; catechin; epicatechin; total phenolic content; Criollo cocoa; kinetic

---

1. Introduction

Cocoa beans, the seeds of the tree *Theobroma cacao* L., are the key raw material for chocolate production [1]. Criollo is the finest variety of cocoa [2]. Its seeds are aromatic, mild tasting, and with low bitterness: they represent the ideal raw material for a quality chocolate [1,3]. For obtaining chocolate from cocoa beans, one very important stage in the process is the roasting [4], which results in the production of desirable flavor and aroma compounds, as well as color changes [5]. The temperature and duration of roasting substantially affected the character in chemical and physical changes of cocoa beans [6].

Cocoa and chocolate have recently gained much attention due to their potential benefits in human health [1,7–10] and have recently become the target of increased scientific research due to their health promoting properties [11]. Cocoa, from a therapeutic viewpoint, is important due to the high concentrations of polyphenols as antioxidants [12]. These health effects have been assumed to be associated with the presence of polyphenols, among which are monomeric flavan-3-ols: epicatechin and catechin [5,13–16]. The beneficial effects of cocoa polyphenols to human health are, among others, the scavenge of free radicals and prevention of damage to DNA, the chelation of metals, vasoprotective effects, improvements in endothelial function, anti-inflammatory effects, the amelioration of insulin resistance, and anticarcinogenic effects, among others [13,17–21]. Cocoa is a rich source of polyphenolic compounds and may account for 12–18% of the dry mass of the beans [5]. Typically, polyphenols are sensitive to heat during the process, especially under a high temperature environment, i.e., roasting and drying [22,23]. Additionally, roasting influences the alteration of bioactive compounds [24]. In particular, the roasting process leads to the loss or modification of flavanols, which leads to a 14% loss of the total phenolic content (TPC), as well as the epimerization of epicatechin to catechin [25–32].

Kinetic modeling can provide a deeper understanding of the changes that occur during thermal processing controlling and food quality optimization [33,34]. Since these roasting techniques were introduced to the chocolate industry, the roasting time and temperature have been studied, applying designs or models that allow for a proper assessment of the process [35]. Various studies have been reported on the effect of roasting on cocoa nibs' coloration, physical/chemical changes [4,5,8,16,23,25,36], flavor changes [37], and the kinetics of polyphenol degradation during drying [12,28,38]; however, studies on the kinetics of the polyphenol and monomeric flavan-3-ol degradation of Criollo cocoa during the roasting process are scarce, so the aim of the present study was to understand this degradation and determine the kinetic parameters.

2. Materials and Methods

2.1. Materials

Approximately 14 kg of dried fermented cocoa beans of Criollo variety were obtained directly from Multi-Service Cooperative APROCAM in the Bagua province, Amazon region, Perú.

2.2. Chemicals and Standards

Methanol HPLC grade (JT Baker, Deventer, The Netherlands), Folin–Ciocalteu's phenol reagent, gallic acid, sodium carbonate, ≥98% (-)-epicatechin (HPLC) from green tea, ≥97% (-)-catechin (HPLC) from green tea, and ≥90% petroleum ether were purchased from Sigma Aldrich (Diessenhofen, Germany).

2.3. Roasting Process

Samples of cocoa bean were subjected to different treatments of time and temperature as conditions of the roasting process. Prior to the roast, the beans were selected according to their size, choosing beans of uniform size. Criollo cocoa samples (100 g) were roasted in a roaster (IMSA, ERTC-51, Lima, Perú) in the Agroindustry Plant Pilot at Universidad Nacional Toribio Rodríguez de Mendoza de Amazonas (UNTRM), Perú.

2.4. Chemical Analysis

2.4.1. Methanolic Extraction of Phenolic Compounds and Monomeric Flavan-3-ols

According to Summa et al. [29] regarding some modifications, three cocoa beans were ground in a pestle from each sample. The powder was defatted by extraction with petroleum ether and centrifuged using a centrifuge (MPW Med Instruments, MPW-51, Warszawa, Poland) at 3000 rpm for 15 min at room temperature. The supernatant was then discarded. Fresh petroleum ether was added and then centrifuged (four times). The resulting defatted material was air dried at room temperature. Methanol extraction based on the methodology used by Jonfia-Essien et al. [30] with some modifications was performed, and 0.5 g of defatted cocoa powder was homogenized in 25 mL of 80% methanol for 30 min in a magnetic stirrer and then filtered in a vacuum filter. This methanolic extract was used for the determination of both TPC and monomeric flavan-3-ols.

2.4.2. Total Phenolic Content

The TPC of the Criollo cocoa beans was determined following the procedures of Singleton et al. [31] and Hu et al. [7]. Diluted extract (0.1 mL) or blank (0.1 mL deionized water) was mixed with 7.9 mL water and 0.5 mL of Folin-Ciocalteu reagent for 5 min at 22 °C. Next, 1.5 mL of saturated sodium carbonate solution was added. Reagents were mixed thoroughly by vigorous shaking for 10 s by hand. The mixture was incubated in the dark at 22 °C for 2 h before determination of the absorbance at 765 nm using an UV/Visible spectrophotometer (Unico, S2100, Dayton, NJ, USA). Gallic acid in 70% methanol was diluted (2–16 mg/L) to create a calibration curve. TPC is expressed as mg of gallic acid equivalents/g defatted cocoa bean (mg GAE/gdf).

2.4.3. Quantification of Epicatechin and Catechin of the Methanolic Extract

Quantification of epicatechin and catechin followed the procedures of Wang et al. [39] with some modifications in a high performance liquid chromatography system (HPLC) (Lachrom Elite, Hitachi, Japan) equipped with a UV-Vis detector (Hitachi, Lachrom Elite L–2420, Wako, Japan) and isocratic pump (Hitachi, Lachrom Elite L–2130, Japan). The column used was a C18 150 × 4.6 mm, 5 μm (Merck, Purospher RP–18 endcapped, Darmstadt, Germany). The mobile phase was methanol/water/orthophosphoric acid (20/79.9/0.1), and the flow rate was 1 mL/min. Absorption wavelength was selected at 210 nm. The sample injection volume was 20 μL. Chromatographic peaks in the samples were identified by comparing their retention time and UV spectrum with those of the reference standards. A standard graph for each component was prepared by plotting concentration versus area. Quantification was carried out from integrated peak areas of the sample and corresponding standard graphs, and the epic/cat ratios of epicatechin and catechin concentrations were obtained.

2.4.4. TPC and Monomeric Flavan-3-ol Degradation

According to Martins et al. [40], the degradation was then calculated according to the following formula:

$$\% \ degradation = \left(\frac{X_{control} - X_t}{X_{control}} \right) * 100 \tag{1}$$

where $X_{control}$ is the TPC, epicatechin, or catechin concentration in the control sample (unroasted cocoa bean), and X_t is the TPC, epicatechin, or catechin concentration at time t.

2.5. Experimental Desing for Kinetics of TPC and Monomeric Flavan-3-ols

Samples (100 g) in triplicate of Criollo cocoa beans were roasted for each combination of time and temperature (10, 20, 30, 40, and 50 min at 90, 110, 130, 150, 170, 190, and 200 °C), 105 samples were obtained. Means and standard deviations were calculated. The temperature and time values used were defined according to the published literature. Fitting procedures (Matlab 2014) were used to determine the reaction rate constants for TPC, epicatechin, and catechin. The general kinetic models of zero-, first-, and second-order reactions were used, presented in Equations (2)–(4), respectively [41].

$$[A]_0 - [A] = kt \tag{2}$$

$$[A] = [A]_0 \ exp(-kt) \tag{3}$$

$$\frac{1}{[A]} - \frac{1}{[A]_0} = kt \tag{4}$$

k (min^{-1}) is the reaction rate constant at temperature T; t is the reaction time (min); $[A]_0$ and $[A]$ are the initial (control sample) and final amounts of TPC, epicatechin or catechin, respectively, at different times t and temperatures.

The final fitting models were obtained by matching and analyzing the initial models obtained by Matlab software with general equations of zero-, first-, and second-order reactions presented in Equations (2)–(4), respectively. The effect of temperature was evaluated by means of the Arrhenius equation.

$$k = k_0 * e^{(-E_a/RT)}. \tag{5}$$

The activation energy E_a for the formation of each parameter was determined by linear regression of $Ln\,k$ curve versus $1/T$ with Equation (6).

$$Lnk = Lnk_0 - E_a/RT \tag{6}$$

E_a (J/mol) is the apparent activation energy, R (8.3145 J/mol.K) is the universal gas constant, k is the reaction rate constant, and k_0 is the pre-exponential factor.

2.6. Statistical Analysis

The results were compared using one-factor analysis of variance (ANOVA) followed by the Tukey test. Previously, Dixon's Q test, for the identification and rejection of outliers, was used. Statistical analyses were carried out in Minitab 17 software.

3. Results

3.1. Effect of Roasting on Monomeric Flavan-3-ols and TPC

The TPC, epicatechin, and catechin concentration and the epi/cat ratio of the unroasted (control sample) Criollo cocoa beans were 110.98 ± 1.43 mg GAE/gdf, 30.29 ± 1.0 mg/gdf, 2.71 ± 0.13 mg/gdf, and 11.20 ± 0.40, respectively (Table 1). These values served as a starting point for the study of degradation. The TPC and epicatechin values and the epi/cat ratios of the treatments were lower than the control sample. Some results show that the concentration of epicatechin in some cases increased. During the roasting process, the epi/cat ratio was reduced as the process temperature increased. Considering a time of 10 min, the epi/cat ratio was reduced from 6.08 ± 0.96 when heated at 90 °C to 3.20 ± 0.68 at 200 °C. Table 1 shows that epicatechin degradation reached 100% when Criollo cocoa beans were roasted for 50 min at 190 or 200 °C, while TPC did not reach full degradation at any temperature or time. The opposite happened with catechin, which showed patterns of formation. The values in bold indicate that there was formation instead of degradation.

Table 1. Concentration and degradation of TPC and monomeric flavan-3-ols, epi/cat ratios, and the formation of catechin.

T (°C)	Time (min)	Concentration [1]			Epi/Cat Ratio [1]	Degradation [1] (%)		
		TPC (mg GAE/gdf)	Epicatechin (mg/gdf)	Catechin (mg/gdf)		TPC	Epicatechin	Catechin [2]
Control	0	110.98 ± 1.43^a	30.29 ± 1.09^a	2.71 ± 0.13^a	11.20 ± 0.40^a			
90	10	54.60 ± 10.86^b	14.97 ± 1.45^b	2.48 ± 0.14^{ab}	6.08 ± 0.96^b	50.84 ± 9.47^a	50.41 ± 6.61^a	8.31 ± 5.43^a
	20	72.70 ± 7.44^b	17.78 ± 3.13^b	2.48 ± 0.05^{ab}	7.20 ± 1.37^b	34.44 ± 7.46^a	41.25 ± 10.23^a	8.31 ± 6.24^a
	30	45.10 ± 7.61^b	11.78 ± 1.22^b	2.20 ± 0.11^b	5.35 ± 0.32^b	59.41 ± 6.47^a	61.07 ± 4.13^a	18.69 ± 3.99^a
	40	62.10 ± 19.80^b	12.92 ± 4.89^b	2.22 ± 0.28^b	5.93 ± 2.62^b	43.90 ± 18.7^a	57.0 ± 17.7^a	18.25 ± 6.22^a
	50	51.58 ± 7.44^b	13.75 ± 2.92^b	2.48 ± 0.04^{ab}	5.55 ± 1.24^b	53.46 ± 7.29^a	54.78 ± 8.05^a	8.08 ± 5.82^a
110	10	68.54 ± 4.05^c	27.24 ± 0.53^c	2.60 ± 0.07^{cd}	10.50 ± 0.36^c	38.21 ± 4.44^c	10.02 ± 2.08^d	2.22 ± 1.45^d
	20	60.84 ± 5.59^{cd}	15.36 ± 0.47^d	2.67 ± 0.21^c	5.78 ± 0.50^d	45.14 ± 5.72^{bc}	49.20 ± 3.07^c	7.78 ± 2.00^{cd}
	30	52.25 ± 3.86^d	11.47 ± 0.46^e	2.28 ± 0.08^{de}	5.05 ± 0.28^{de}	52.91 ± 3.59^b	62.10 ± 1.50^b	15.79 ± 4.65^{bc}
	40	48.14 ± 6.12^d	8.28 ± 1.36^e	2.33 ± 0.18^{cde}	3.54 ± 0.31^f	56.58 ± 5.95^b	72.53 ± 5.49^b	13.59 ± 8.53^{bcd}
	50	58.92 ± 6.57^{cd}	8.22 ± 2.37^e	2.06 ± 0.12^e	3.96 ± 0.97^{ef}	46.88 ± 6.30^{bc}	72.77 ± 8.31^b	24.00 ± 3.58^b
130	10	98.85 ± 7.79^e	28.74 ± 0.94^f	3.01 ± 0.03^f	9.55 ± 0.39^g	10.98 ± 6.04^f	8.05 ± 3.01^e	**11.46 ± 4.71^e**
	20	52.24 ± 3.68^f	$11.21 \pm 1.24g$	2.69 ± 0.47^f	4.20 ± 0.26^h	52.93 ± 3.24^e	62.87 ± 5.25^d	14.22 ± 6.08^e
	30	35.34 ± 2.19^g	9.47 ± 1.01^{gh}	2.62 ± 0.23^f	3.66 ± 0.67^{hi}	68.14 ± 2.34^d	68.77 ± 2.57^{cd}	5.30 ± 4.23^e
	40	44.35 ± 7.73^{fg}	10.46 ± 1.46^g	3.30 ± 0.36^f	3.23 ± 0.81^{hi}	60.01 ± 7.15^{de}	65.40 ± 5.08^d	**18.62 ± 7.50^e**
	50	38.43 ± 3.98^{fg}	6.98 ± 1.01^h	2.95 ± 0.22^f	2.36 ± 0.20^i	65.40 ± 3.16^{de}	76.99 ± 2.78^c	**9.08 ± 7.50^e**
150	10	46.71 ± 7.84^h	11.25 ± 2.11^i	2.62 ± 0.24^g	4.26 ± 0.41^j	57.85 ± 7.61^h	62.70 ± 8.19^h	8.66 ± 0.15^f
	20	48.78 ± 8.12^h	8.70 ± 0.70^{ij}	3.49 ± 0.68^g	2.59 ± 0.73^k	56.07 ± 7.02^h	71.29 ± 1.29^{gh}	**32.90 ± 25.60^f**
	30	31.32 ± 2.25^i	6.32 ± 0.53^{jk}	3.79 ± 0.75^g	1.70 ± 0.25^{kl}	71.80 ± 1.72^g	79.14 ± 1.51^{fg}	**39.60 ± 21.10^f**
	40	31.27 ± 2.17^i	5.72 ± 0.12^k	3.64 ± 0.10^g	1.57 ± 0.07^{kl}	71.81 ± 2.18^g	81.11 ± 0.84^{fg}	**34.61 ± 8.41^f**
	50	28.22 ± 1.32^i	4.27 ± 0.68^k	3.01 ± 0.35^g	1.41 ± 0.08^l	74.57 ± 1.37^g	85.95 ± 1.81^f	**11.05 ± 8.82^f**
170	10	42.10 ± 7.56^j	12.33 ± 2.33^l	3.48 ± 0.98^{hij}	0.08 ± 0.02^{no}	62.12 ± 6.39^j	59.44 ± 6.36^k	**11.58 ± 0.01^h**
	20	33.43 ± 7.28^{jk}	6.75 ± 0.77^m	4.66 ± 0.31^{hi}	0.10 ± 0.02^n	69.84 ± 6.81^{ij}	77.75 ± 1.80^j	**72.34 ± 12.07^{gh}**
	30	31.50 ± 6.43^{jk}	5.41 ± 0.69^{mn}	5.15 ± 1.11^h	0.17 ± 0.05^m	71.66 ± 5.50^{ij}	82.11 ± 2.36^{ij}	**84.90 ± 37.90^g**
	40	31.33 ± 7.40^{jk}	4.16 ± 0.59^{mn}	3.85 ± 0.66^{hij}	0.12 ± 0.02^{mn}	71.82 ± 6.36^{ij}	86.26 ± 1.77^{ij}	**43.00 ± 30.30^{gh}**
	50	19.54 ± 1.05^k	3.12 ± 0.18^n	3.19 ± 0.42^{ij}	0.11 ± 0.01^{mn}	82.40 ± 0.89^i	89.70 ± 0.77^i	**22.19 ± 14.41^h**
190	10	41.13 ± 8.69^l	11.61 ± 4.88^o	3.67 ± 0.91^l	3.35 ± 1.62^p	62.95 ± 7.70^l	61.93 ± 15.42^m	42.3 ± 29.40^{ij}
	20	38.17 ± 6.69^l	5.70 ± 0.50^p	4.87 ± 0.28^k	1.18 ± 0.17^q	65.63 ± 5.75^l	81.20 ± 0.99^l	**80.50 ± 17.40^i**
	30	21.35 ± 4.45^m	3.11 ± 0.60^{pq}	2.77 ± 0.25^{lm}	1.11 ± 0.12^q	80.79 ± 3.80^k	89.70 ± 2.32^{kl}	7.32 ± 6.32^j
	40	19.81 ± 2.11^m	2.43 ± 0.13^{pq}	2.67 ± 0.27^{lm}	0.92 ± 0.10^q	82.17 ± 1.72^k	91.97 ± 0.23^{kl}	9.97 ± 3.21^j
	50	12.31 ± 1.36^m	0.00 ± 0.00^q	2.05 ± 0.07^m	0.00 ± 0.00^q	88.91 ± 1.27^k	100.00 ± 0.00^k	23.96 ± 6.02^j
200	10	33.23 ± 2.44^n	8.10 ± 0.88^r	3.20 ± 0.68^n	2.64 ± 0.83^r	70.04 ± 2.60^o	73.21 ± 3.58^p	**21.89 ± 14.12^k**
	20	17.33 ± 3.26^o	2.42 ± 0.21^s	2.30 ± 0.30^n	1.06 ± 0.05^s	84.38 ± 2.98^n	92.01 ± 0.39^o	15.17 ± 7.86^k
	30	13.76 ± 0.53^{op}	2.32 ± 0.06^s	2.26 ± 0.07^n	1.03 ± 0.03^s	87.60 ± 0.64^{mn}	92.34 ± 0.10^o	16.56 ± 1.54^k
	40	14.41 ± 1.86^o	2.53 ± 0.45^s	2.35 ± 0.60^n	1.09 ± 0.09^{st}	87.00 ± 1.85^{mn}	91.61 ± 1.81^o	22.29 ± 10.34^k
	50	8.56 ± 0.10^p	0.00 ± 0.00^t	2.13 ± 0.21^n	0.00 ± 0.00^t	92.29 ± 0.06^m	100.00 ± 0.00^n	21.37 ± 5.74^k

[1] Different letters in the same column represent statistically significant differences ($p \leq 0.05$). At least three replicate samples were analyzed. [2] Bold values indicate an increase in catechin concentration.

3.2. Roasting Kinetics of Monomeric Flavan-3-ols and TPC

The values obtained from the fitted parameters are given in Table 2. In the case of TPC and epicatechin, the k value increases from 0.02 ± 0.01 min^{-1} at 90 °C to 0.10 ± 0.05 min^{-1} at 200 °C, and from 0.02 ± 0.01 min^{-1} at 90 °C to 0.13 ± 0.04 min^{-1} at 200 °C, respectively. The kinetic parameters of the catechin are not shown because it corresponds to a combined production and degradation model, which does not correspond to Equations (2), (3), or (4).

Table 2. First-order kinetic parameters fitted for TPC and epicatechin degradation.

Roasting Temperature (°C)	TPC			Epicatechin		
	k (min^{-1})	R^2	RMSE	k (min^{-1})	R^2	RMSE
90	0.02 ± 0.01	0.52	18.50	0.02 ± 0.01	0.65	4.57
110	0.02 ± 0.01	0.70	13.97	0.03 ± 0.01	0.95	2.49
130	0.03 ± 0.02	0.87	13.08	0.03 ± 0.02	0.86	4.31
150	0.04 ± 0.03	0.83	14.59	0.06 ± 0.03	0.91	3.27
170	0.07 ± 0.03	0.96	2.24	0.07 ± 0.03	0.96	2.24
190	0.06 ± 0.03	0.92	11.35	0.09 ± 0.02	0.99	1.05
200	0.10 ± 0.05	0.96	9.23	0.13 ± 0.04	0.99	1.44

Figure 1 shows that the catechin reaction kinetics could be divided into two order models: one for formation and the other for degradation, where the highest catechin formation (**84.90 ± 37.90%**) was at 170 °C and 30 min of roasting.

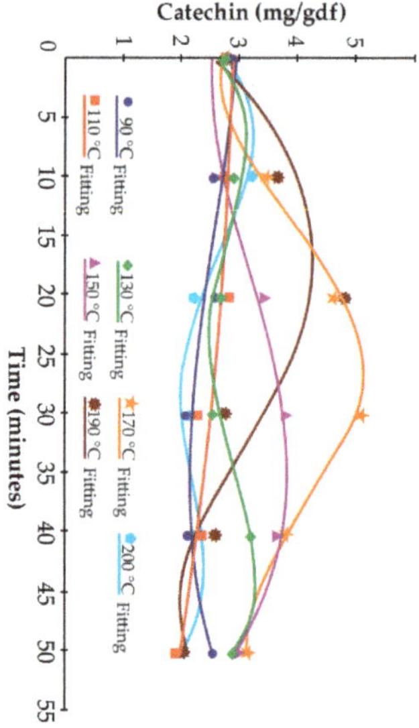

Figure 1. Kinetic of formation and degradation of catechin.

The temperature dependence of the TPC and epicatechin degradation was estimated using the Arrhenius equation expressed in Equation (5). The linear behavior of $Ln\ k$ versus $1/T$ allowed us to determine E_a for TPC and epicatechin of 24.03 J/mol ($R^2 = 0.94$) and 22.51 J/mol ($R^2 = 0.97$), respectively (Figure 2). The parameters for catechin kinetics were not calculated because it presented a degradation and production pattern, as mentioned above.

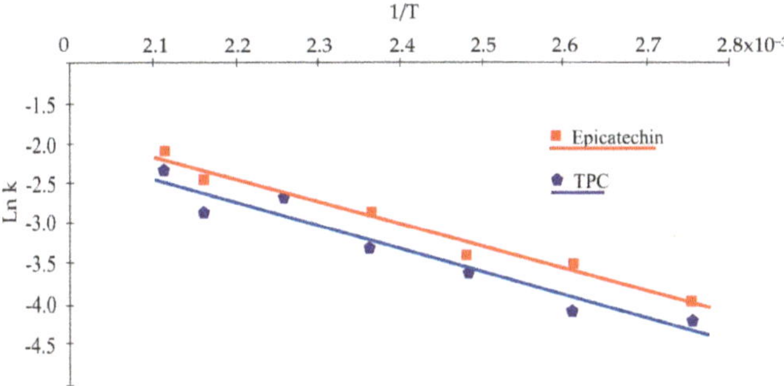

Figure 2. Correlation of *Ln k* with 1/T to obtain the TPC and epicatechin kinetic parameters.

4. Discussion

The time and temperature of the roasting process depend on several factors, such as the type of cocoa (Criollo or Forastero) and others [42]. The main type of polyphenols (known for their demonstrated antioxidant and anti-inflammatory properties) in cocoa is flavanols. This family of compounds includes catechin and epicatechin (monomeric species). Epicatechin is the most abundant flavanol in cocoa and accounts for 35% of the total polyphenolic fraction [43,44] (TPC). In one study made by Kim and Keeney [45], the epicatechin concentrations ranged from 2.66 (Jamaica) to 16.52 mg/g (Costa Rica) of the defatted sample in cocoa beans of different varieties. The epicatechin and catechin concentrations of the control sample were higher than those found by Kim and Keeney [45] and Payne et al. [32] in fermented cocoa beans (the Forastero variety) from Ivory Coast (epicatechin: 1.69 ± 0.10 mg/g; catechin: 0.08 ± 0.00 mg/g) and Papua New Guinea (epicatechin: 0.78 ± 0.04 mg/g; catechin: 0.05 ± 0.00 mg/g). These values were higher than those found by Mazor Jolić et al. [46] in Forastero cocoa beans from Ghana (epicatechin: 2.23 ± 0.6 mg/g; catechin: 0.28 ± 0.04 mg/g). Other studies conducted in Perú on fermented cocoa beans from Tingo María, San Alejandro, and Curimaná presented epicatechin and catechin concentrations of 0.33–5.04 mg/g and 0.02–0.14 mg/g, respectively [47]. It is known that epicatechin has diverse biological properties (antioxidant, antimicrobial, anti-inflammatory, antitumor, and cardio-protective activity) [48]; thus, the Criollo cocoa beans used in the present study show potential for use in the elaboration of functional foods. It has been postulated that the ratio of epicatechin to catechin (epi/cat) possibly could be associated with the degree of cocoa processing [32,49]. Payne et al. [32] obtained epi/cat ratio values of 20.1 ± 0.63 for dry fermented cocoa beans and 3.35 ± 0.20 at 90 °C to 0.96 ± 0.02 at 120 °C for roasted beans. Table 1 shows control sample epi/cat ratio values of 11.20 ± 0.40; considering 50 min of roasting, values of 5.55 ± 1.24 at 90 °C to 2.36 ± 0.20 at 130 °C and 0.00 ± 0.00 at 200 °C were obtained. These results demonstrate that roasting at 200 °C at any time (10 or 50 min) is aggressive for epicatechin, so if these parameters are used, the functional properties of chocolate will be lost.

In Table 1, a roasting time of 50 min at 90, 110, 130, 150, 170, 190, and 200 °C degraded the TPC to 53.46 ± 7.29, 46.88 ± 6.30, 65.40 ± 3.16, 74.57 ± 1.37, 82.40 ± 0.89, 88.91 ± 1.27, and 92.29 ± 0.06, respectively; these results are consistent with those obtained by Mazor Jolić et al. [46]. The decrease of TPC is associated with the thermal and oxidative degradation of these compounds [35]. It was proven that, at high temperature, low molecular weight phenolic compounds easily volatilize [50]. Djikeng et al. [50] also observed that the degradation percentage of the TPC also increased with roasting time. In general, more intense roasting conditions result in a greater loss of TPC due to the high redox activity of polyphenols at those conditions [50,51]. Epicatechin degraded more than TPC at a higher temperature, obtaining a complete degradation at 200 °C and 50 min. This is concordant with Stanley et al. [5], who observed the significant effects of roasting time within temperatures up to 190 °C; the levels of epicatechin in Trinitario cocoa (a type of fine cocoa) decreased in a time- and temperature-dependent manner. Significant decreases in Criollo cocoa were observed. Among the monomeric flavan-3-ols, epicatechin was identified as the more active compound responsible for the vascular health benefits associated with cocoa and chocolate [52]. Applying treatments of 190–200 °C results in a total loss of epicatechin, and the healthy properties of the Criollo cacao and its aromatic properties may also be reduced; for this reason, Żyżelewicz et al. [1] and Hurst et al. [36] state that Criollo beans require milder roasting conditions. The roasting temperature is generally in the range from 110 to 120 °C.

The most important reactions occurring with catechins under thermal processing are epimerization, hydrolysis, oxidation, and polymerization; catechin epimerization takes place on two asymmetric carbon atoms in the C ring [53]. In Table 1, catechin shows a particular behavior due to the increase in its concentration (formation) at the beginning of the roasting. The catechin values in bold font shows an increase in its concentration with respect to the control sample (unroasted beans) during the first 10 min at 130 °C (**11.46 ± 4.71%**), and its formation continues until the first 10 min at 200 °C and subsequently degrades to 21.37 ± 5.74%. The trend of these results is consistent with those obtained by Stanley et al. [5], except that the catechin formation was much higher, showing an increase in catechin concentration 675% higher than the control (unroasted beans) at 10 min. These results were reported for Trinitario cocoa beans. We can affirm that, in Trinitario cacao, the formation of catechin is much greater than in Criollo cacao, and this applies to the epimerization of epicatechin to catechin [5,14,25,32,54]; however, as the treatment continued, its concentration decreased. These results were also found by Żyżelewicz et al. [4] in Forastero cocoa, since after 15 min of roasting at 135 °C, the content of catechin increased significantly, initiating its degradation. When cocoa beans are progressively roasted at conditions described as low, medium, and high roast conditions (160 °C at 13, 20, and 25 min), there is a progressive loss of epicatechin and an increase in catechin with higher roast levels [36,55]. The high temperature–short time (HTST) process induces higher epicatechin epimerization than does the low temperature–long-time (LTLT) process, generating greater amounts of catechin [16,53].

Cocoa consumption is suggested to promote health benefits. The amounts and profiles of monomeric flavanols depend strongly on the bean type, origin, and manufacturing process. Roasting is known as a crucial step in the technical treatment of cocoa, which leads to flavanol losses and modifications, especially the epimerization of epicatechin to catechin [42]. These modifications were fully produced (epimerization) at temperatures of 150, 170, and 190 °C (Table 1). In the manufacture of chocolate, it is necessary to achieve a balance that maintains healthy properties and the development of the aroma that characterizes fine, Criollo cocoa chocolates.

In order to explain the phenomena of TPC and epicatechin degradation, the data were fitted using kinetic models, which are pseudo first-order kinetic reactions coinciding with the majority of food reactions [56]. The term "pseudo" is added with respect to reactions in biological materials, such as food, because the actual reaction mechanism and its kinetics are far more complex [28]. We used seven temperatures in our experiment, allowing us to notice the behavior of the roasting process; when Taoukis et al. [56] states that five or six experimental temperatures are the practical optimum to obtain meaningfully narrow confidence limits in kinetic parameters (E_a and k). The values obtained from the fitted parameters in Matlab are given in Table 2 and consistent with previous studies [26,28,41,57], in which the first-order kinetic model explains both TPC and epicatechin degradation at all temperatures; the k value of epicatechin is higher than the k value of TPC. These results suggest that the degradation of epicatechin is faster than the degradation of TPC. Roasting conditions of 130 °C and 10 min produced a lower percentage of epicatechin (8.05 ± 3.01%) and TPC (10.98 ± 6.04%) degradation. Many authors claim that these compounds are degraded by heat [4,22,23,50]. The most intense temperatures have caused significant degradation of the components; however, it was found that the epicatechin is more thermosensitive than TPC. This is explained by the greater k value that epicatechin has compared to TPC; however, the chemical basis of this phenomenon must be studied.

Kinetic modeling may also use the influence of processing on critical quality parameters. Knowledge of degradation kinetics, including the reaction order, E_a and k, is of great importance to predict food quality loss during thermal process treatments [58]. Food quality loss reactions described by the kinetic models were shown to follow Arrhenius (Equation (5)) behavior with temperature changes [56]. The E_a for TPC was greater than the E_a for epicatechin, this being the minimal value of energy that a specific collision between reagent molecules must achieve in order for a reaction to take place [59]. The epicatechin thus needs less energy than TPC to degrade it. The relative differences in E_a values could be due to the different composition of the sample studied or to changes occurring in the samples during heating [58]. Olivares-Tenorio et al. [60] observed that catechin followed patterns of formation by epimerization and degradation, and experiments need to be devised to unravel these various reactions. Figure 1 shows that catechin reaction kinetics could be divided into a two order models: one for formation and the other for degradation, where the highest catechin formation was at 170 °C and 30 min (**84.90 ± 37.90%**), followed by treatments at 190 °C and 150 °C. The largest catechin formation occurs in the first 30 min of roasting at any of the aforementioned temperatures. The epimerization from epicatechin to catechin due to technological treatment (heat) has often been postulated. Only a few publications have confirmed this reaction by enantioseparation. The reaction mechanism is not fully clarified, but it is assumed that ring opening occurs at position C-2 of the oxygenated ring, and reclosing leads to the atypical enantiomers. High temperatures, particularly when combined with alkaline conditions, accelerate the epimerization reaction [42]. Figure 1 shows that at 90 and 110 °C there is only catechin degradation; subsequently, the formation of catechin begins slowly from 130 °C, reaching its highest point at 190 °C. The treatment at 200 °C only shows a degradation pattern. It is assumed that the k for catechin formation kinetics will be greater at 190 °C for 30 min, after which the degradation kinetics will take place at a high k. An enantoseparation reaction will occur to a greater extent for 30 min when the Criollo cocoa beans are roasted at 150, 170, or 190 °C.

5. Conclusions

The roasting process of Criollo cocoa allowed for a higher formation of catechin in the first 30 min of the process at 170 °C, followed by degradation to minimum levels at 200 °C, while the epicatechin showed greater susceptibility to treatment than the TPC and the catechin. Likewise, the degradation data of TPC and epicatechin are better suited to a first-order kinetic reaction as the temperature increases. Although it is true that the roasting stage is essential for the development of the aromas of chocolate, this also affects the polyphenolic content and monomeric flavan-3-ol of cocoa beans and thus the final chocolate product; therefore, roasting at moderate temperatures is necessary to obtain minimal degradation of cocoa phenolic compounds and consequently antioxidant properties. Likewise, with epicatechin being part of the total phenolic compounds, its degradation percentage is higher. This is explained by its kinetics; however, its chemical explanation for future work is pending.

Author Contributions: Conceptualization, E.M.C.-A. and F.P.C.-T.; methodology, E.F.-R. and S.G.C.-Q.; software, E.M.C.-A.; formal analysis, E.M.C.-A. and F.P.C.-T.; investigation, E.F.-R.; writing—original draft preparation, E.F.-R.; writing—review and editing, E.M.C.-A., R.S., and F.P.C.-T.; supervision, F.P.C.-T. All authors have read and agreed to the published version of the manuscript.

Funding: This research was funded by Project N°16808-2016/PROYECTO-CACAO, of the Programa Nacional de Innovación Agraria (PNIA) of the Instituto Nacional de Innovación Agraria of the Peruvian Government, the Universidad Nacional Toribio Rodríguez de Mendoza de Amazonas, and the Pontificia Universidad Católica del Perú.

Acknowledgments: A special acknowledgement to Multi-Service Cooperative APROCAM for the permission for the use of their facilities.

Conflicts of Interest: The authors declare that there is no conflict of interest.

References

1. Żyżelewicz, D.; Budryn, G.; Oracz, J.; Antolak, H.; Kręgiel, D.; Kaczmarska, M. The effect on bioactive components and characteristics of chocolate by functionalization with raw cocoa beans. *Food Res. Int.* **2018**, *113*, 234–244. [CrossRef] [PubMed]
2. Castro-Alayo, E.M.; Idrogo-Vásquez, G.; Siche, R.; Cardenas-Toro, F.P. Formation of aromatic compounds precursors during fermentation of Criollo and Forastero cocoa. *Heliyon* **2019**, *5*, e01157. [CrossRef]

3. Ascrizzi, R.; Flamini, G.; Tessieri, C.; Pistelli, L. From the raw seed to chocolate: Volatile profile of Blanco de Criollo in different phases of the processing chain. *Microchem. J.* **2017**, *133*, 474–479. [CrossRef]

4. Żyżelewicz, D.; Krysiak, W.; Oracz, J.; Sosnowska, D.; Budryn, G.; Nebesny, E. The influence of the roasting process conditions on the polyphenol content in cocoa beans, nibs and chocolates. *Food Res. Int.* **2016**, *89*, 918–929. [CrossRef]

5. Stanley, T.H.; Van Buiten, C.B.; Baker, S.A.; Elias, R.J.; Anantheswaran, R.C.; Lambert, J.D. Impact of roasting on the flavan-3-ol composition, sensory-related chemistry, and in vitro pancreatic lipase inhibitory activity of cocoa beans. *Food Chem.* **2018**, *255*, 414–420. [CrossRef] [PubMed]

6. Oracz, J.; Nebesny, E. Influence of roasting conditions on the biogenic amine content in cocoa beans of different Theobroma cacao cultivars. *Food Res. Int.* **2014**, *55*, 1–10. [CrossRef]

7. Hu, Y.; Pan, Z.J.; Liao, W.; Li, J.; Gruget, P.; Kitts, D.D.; Lu, X. Determination of antioxidant capacity and phenolic content of chocolate by attenuated total reflectance-Fourier transformed-infrared spectroscopy. *Food Chem.* **2016**, *202*, 254–261. [CrossRef]

8. Sacchetti, G.; Ioannone, F.; De Gregorio, M.; Di Mattia, C.; Serafini, M.; Mastrocola, D. Non enzymatic browning during cocoa roasting as affected by processing time and temperature. *J. Food Eng.* **2016**, *169*, 44–52. [CrossRef]

9. Nascimento, M.M.; Santos, H.M.; Coutinho, J.P.; Lôbo, I.P.; da Silva Junior, A.L.S.; Santos, A.G.; de Jesus, R.M. Optimization of chromatographic separation and classification of artisanal and fine chocolate based on its bioactive compound content through multivariate statistical techniques. *Microchem. J.* **2020**, *152*, 104342. [CrossRef]

10. Steinberg, F.M.; Bearden, M.M.; Keen, C.L. Cocoa and chocolate flavonoids: Implications for cardiovascular health. *J. Am. Diet. Assoc.* **2003**, *103*, 215–223. [CrossRef]

11. Do Carmo Brito, B.D.N.; Campos Chisté, R.; Da Silva Pena, R.; Abreu Gloria, M.B.; Santos Lopes, A. Bioactive amines and phenolic compounds in cocoa beans are affected by fermentation. *Food Chem.* **2017**, *228*, 484–490. [CrossRef] [PubMed]

12. Alean, J.; Chejne, F.; Rojano, B. Degradation of polyphenols during the cocoa drying process. *J. Food Eng.* **2016**, *189*, 99–105. [CrossRef]

13. Wang, Y.; Feltham, B.A.; Suh, M.; Jones, P.J.H. Cocoa flavanols and blood pressure reduction: Is there enough evidence to support a health claim in the United States? *Trends Food Sci. Technol.* **2019**, *83*, 203–210. [CrossRef]

14. Alañón, M.E.; Castle, S.M.; Siswanto, P.J.; Cifuentes-Gómez, T.; Spencer, J.P.E. Assessment of flavanol stereoisomers and caffeine and theobromine content in commercial chocolates. *Food Chem.* **2016**, *208*, 177–184. [CrossRef] [PubMed]

15. Fayeulle, N.; Vallverdu-Queralt, A.; Meudec, E.; Hue, C.; Boulanger, R.; Cheynier, V.; Sommerer, N. Characterization of new flavan-3-ol derivatives in fermented cocoa beans. *Food Chem.* **2018**, *259*, 207–212. [CrossRef] [PubMed]

16. Ioannone, F.; Di Mattia, C.D.; De Gregorio, M.; Sergi, M.; Serafini, M.; Sacchetti, G. Flavanols, proanthocyanidins and antioxidant activity changes during cocoa (*Theobroma cacao* L.) roasting as affected by temperature and time of processing. *Food Chem.* **2015**, *174*, 256–262. [CrossRef]

17. Krysiak, W. Influence of roasting conditions on coloration of roasted cocoa beans. *J. Food Eng.* **2006**, *77*, 449–453. [CrossRef]

18. Andres-Lacueva, C.; Monagas, M.; Khan, N.; Izquierdo-Pulido, M.; Urpi-Sarda, M.; Permanyer, J.; Lamuela-Raventós, R.M. Flavanol and Flavonol Contents of Cocoa Powder Products: Influence of the Manufacturing Process. *J. Agric. Food Chem.* **2008**, *56*, 3111–3117. [CrossRef]

19. Martín, M.A.; Ramos, S. Cocoa polyphenols in oxidative stress: Potential health implications. *J. Funct. Foods* **2016**, *27*, 570–588. [CrossRef]

20. Martín, M.Á.; Ramos, S. Health beneficial effects of cocoa phenolic compounds: A mini-review. *Curr. Opin. Food Sci.* **2017**, *14*, 20–25. [CrossRef]

21. Della Pelle, F.; Blandón-Naranjo, L.; Alzate, M.; Del Carlo, M.; Compagnone, D. Cocoa powder and catechins as natural mediators to modify carbon-black based screen-printed electrodes. Application to free and total glutathione detection in blood. *Talanta* **2020**, *207*, 120349. [CrossRef] [PubMed]

22. Hii, C.L.; Menon, A.S.; Chiang, C.L.; Sharif, S. Kinetics of hot air roasting of cocoa nibs and product quality. *J. Food Process Eng.* **2017**, *40*, e12467. [CrossRef]

23. Afoakwa, E. Roasting effects on phenolic content and free radical scavening actiivities of pulp preconditioned and fermented (Theobroma cacao) beans. *Afr. J. Food Agric. Nutr. Dev.* **2015**, *15*, 9635–9650.

24. Van Durme, J.; Ingels, I.; De Winne, A. Inline roasting hyphenated with gas chromatography–mass spectrometry as an innovative approach for assessment of cocoa fermentation quality and aroma formation potential. *Food Chem.* **2016**, *205*, 66–72. [CrossRef]

25. Hu, S.; Kim, B.-Y.; Baik, M.-Y. Physicochemical properties and antioxidant capacity of raw, roasted and puffed cacao beans. *Food Chem.* **2016**, *194*, 1089–1094. [CrossRef]

26. Teh, Q.T.M.; Tan, G.L.Y.; Loo, S.M.; Azhar, F.Z.; Menon, A.S.; Hii, C.L. The Drying Kinetics and Polyphenol Degradation of Cocoa Beans: Cocoa Drying and Polyphenol Degradation. *J. Food Process Eng.* **2016**, *39*, 484–491. [CrossRef]

27. Hii, C.L.; Law, C.L.; Cloke, M.; Suzannah, S. Thin layer drying kinetics of cocoa and dried product quality. *Biosyst. Eng.* **2009**, *102*, 153–161. [CrossRef]

28. Kyi, T.M.; Daud, W.R.W.; Mohammad, A.B.; Wahid Samsudin, M.; Kadhum, A.A.H.; Talib, M.Z.M. The kinetics of polyphenol degradation during the drying of Malaysian cocoa beans. *Int. J. Food Sci. Technol.* **2005**, *40*, 323–331. [CrossRef]

29. Summa, C.; Raposo, F.C.; McCourt, J.; Scalzo, R.L.; Wagner, K.-H.; Elmadfa, I.; Anklam, E. Effect of roasting on the radical scavenging activity of cocoa beans. *Eur. Food Res. Technol.* **2006**, *222*, 368–375. [CrossRef]

30. Jonfia-Essien, W.A.; West, G.; Alderson, P.G.; Tucker, G. Phenolic content and antioxidant capacity of hybrid variety cocoa beans. *Food Chem.* **2008**, *108*, 1155–1159. [CrossRef]

31. Singleton, V.; Orthofer, R.; Lamuela-Raventos, R. Analysis of total phenols and other oxidation substrates and antioxidants by means of Folin Ciocalteu reagent. In *Methods in Enzymology*; Academic Press: Cambridge, UK, 1999; pp. 152–178.

32. Payne, M.J.; Hurst, W.J.; Miller, K.B.; Rank, C.; Stuart, D.A. Impact of Fermentation, Drying, Roasting, and Dutch Processing on Epicatechin and Catechin Content of Cacao Beans and Cocoa Ingredients. *J. Agric. Food Chem.* **2010**, *58*, 10518–10527. [CrossRef]

33. Van Boekel, M.A.J.S.; Tijskens, L.M.M. Kinetic modelling. In *Food Process Modelling*; Woodhead Publishing Limited: Cambridge, UK, 2001; pp. 35–59. ISBN 978-1-85573-565-1.

34. Rabeler, F.; Feyissa, A.H. Kinetic Modeling of Texture and Color Changes During Thermal Treatment of Chicken Breast Meat. *Food Bioprocess Technol.* **2018**, *11*, 1495–1504. [CrossRef]

35. García-Alamilla, P.; Lagunes-Gálvez, L.M.; Barajas-Fernández, J.; García-Alamilla, R. Physicochemical Changes of Cocoa Beans during Roasting Process. *J. Food Qual.* **2017**, *2017*, 1–11. [CrossRef]

36. Hurst, W.J.; Krake, S.H.; Bergmeier, S.C.; Payne, M.J.; Miller, K.B.; Stuart, D.A. Impact of fermentation, drying, roasting and Dutch processing on flavan-3-ol stereochemistry in cacao beans and cocoa ingredients. *Chem. Cent. J.* **2011**, *5*, 53. [CrossRef] [PubMed]

37. Baghdadi, Y.M.; Hii, C.L. Mass transfer kinetics and effective diffusivities during cocoa roasting. *J. Eng. Sci. Technol.* **2017**, *12*, 127–137.

38. Menon, A.S.; Hii, C.L.; Law, C.L.; Shariff, S.; Djaeni, M. Effects of drying on the production of polyphenol-rich cocoa beans. *Dry. Technol.* **2017**, *35*, 1799–1806. [CrossRef]

39. Wang, H.; Helliwell, K.; You, X. Isocratic elution system for the determination of catechins, caffeine and gallic acid in green tea using HPLC. *Food Chem.* **2000**, *68*, 115–121. [CrossRef]

40. Martins, L.M.; Sant'Ana, A.S.; Iamanaka, B.T.; Berto, M.I.; Pitt, J.I.; Taniwaki, M.H. Kinetics of aflatoxin degradation during peanut roasting. *Food Res. Int.* **2017**, *97*, 178–183. [CrossRef]

41. Molaveisi, M.; Beigbabaei, A.; Akbari, E.; Noghabi, M.S.; Mohamadi, M. Kinetics of temperature effect on antioxidant activity, phenolic compounds and color of Iranian jujube honey. *Heliyon* **2019**, *5*, e01129. [CrossRef]

42. Kothe, L.; Zimmermann, B.F.; Galensa, R. Temperature influences epimerization and composition of flavanol monomers, dimers and trimers during cocoa bean roasting. *Food Chem.* **2013**, *141*, 3656–3663. [CrossRef]

43. Wollgast, J.; Anklam, E. Review on polyphenols in Theobroma cacao: Changes in composition during the manufacture of chocolate and methodology for identification and quantification. *Food Res. Int.* **2000**, *33*, 423–447. [CrossRef]

44. Quelal-Vásconez, M.A.; Lerma-García, M.J.; Pérez-Esteve, É.; Arnau-Bonachera, A.; Barat, J.M.; Talens, P. Changes in methylxanthines and flavanols during cocoa powder processing and their quantification by near-infrared spectroscopy. *LWT* **2020**, *117*, 108598. [CrossRef]

45. Kim, H.; Keeney, P.G. (-)-Epicatechin Content in Fermented and Unfermented Cocoa Beans. *J. Food Sci.* **1984**, *49*, 1090–1092. [CrossRef]

46. Mazor Jolić, S.; Radojčić Redovniković, I.; Marković, K.; Ivanec Šipušić, Đ.; Delonga, K. Changes of phenolic compounds and antioxidant capacity in cocoa beans processing: Changes of phenolic in cocoa beans processing. *Int. J. Food Sci. Technol.* **2011**, *46*, 1793–1800. [CrossRef]

47. Peláez, P.; Bardón, I.; Camasca, P. Methylxanthine and catechin content of fresh and fermented cocoa beans, dried cocoa beans, and cocoa liquor. *Sci. Agropecu.* **2016**, *7*, 355–365. [CrossRef]

48. Prakash, M.; Basavaraj, B.V.; Chidambara Murthy, K.N. Biological functions of epicatechin: Plant cell to human cell health. *J. Funct. Foods* **2019**, *52*, 14–24. [CrossRef]

49. Miller, K.B.; Hurst, W.J.; Flannigan, N.; Ou, B.; Lee, C.Y.; Smith, N.; Stuart, D.A. Survey of Commercially Available Chocolate- and Cocoa-Containing Products in the United States. 2. Comparison of Flavan-3-ol Content with Nonfat Cocoa Solids, Total Polyphenols, and Percent Cacao. *J. Agric. Food Chem.* **2009**, *57*, 9169–9180. [CrossRef]

50. Djikeng, F.T.; Teyomnou, W.T.; Tenyang, N.; Tiencheu, B.; Morfor, A.T.; Touko, B.A.H.; Houketchang, S.N.; Boungo, G.T.; Karuna, M.S.L.; Ngoufack, F.Z.; et al. Effect of traditional and oven roasting on the physicochemical properties of fermented cocoa beans. *Heliyon* **2018**, *4*, e00533. [CrossRef]

51. Suazo, Y.; Davidov-Pardo, G.; Arozarena, I. Effect of Fermentation and Roasting on the Phenolic Concentration and Antioxidant Activity of Cocoa from Nicaragua: Effect of Process on Cocoa from Nicaragua. *J. Food Qual.* **2014**, *37*, 50–56. [CrossRef]

52. Schroeter, H.; Heiss, C.; Balzer, J.; Kleinbongard, P.; Keen, C.L.; Hollenberg, N.K.; Sies, H.; Kwik-Uribe, C.; Schmitz, H.H.; Kelm, M. (-)-Epicatechin mediates beneficial effects of flavanol-rich cocoa on vascular function in humans. *Proc. Natl. Acad. Sci. USA* **2006**, *103*, 1024–1029. [CrossRef]

53. Lončarić, A.; Pablo Lamas, J.; Guerra, E.; Kopjar, M.; Lores, M. Thermal stability of catechin and epicatechin upon disaccharides addition. *Int. J. Food Sci. Technol.* **2018**, *53*, 1195–1202. [CrossRef]

54. Bernatova, I. Biological activities of (−)-epicatechin and (−)-epicatechin-containing foods: Focus on cardiovascular and neuropsychological health. *Biotechnol. Adv.* **2018**, *36*, 666–681. [CrossRef] [PubMed]

55. Oracz, J.; Nebesny, E.; Żyżelewicz, D. Changes in the flavan-3-ols, anthocyanins, and flavanols composition of cocoa beans of different Theobroma cacao L. groups affected by roasting conditions. *Eur. Food Res. Technol.* **2015**, *241*, 663–681. [CrossRef]

56. Taoukis, P.S.; Labuza, T.P.; Saguy, I.S. Kinetics of Food Deterioration and Shelf-Life Prediction. In *Handbook of Food Engineering Practice*; CRC Press: New York, NY, USA, 1997; pp. 366–408.

57. Henríquez, C.; Córdova, A.; Almonacid, S.; Saavedra, J. Kinetic modeling of phenolic compound degradation during drum-drying of apple peel by-products. *J. Food Eng.* **2014**, *143*, 146–153. [CrossRef]

58. Turturică, M.; Stănciuc, N.; Bahrim, G.; Râpeanu, G. Effect of thermal treatment on phenolic compounds from plum (prunus domestica) extracts—A kinetic study. *J. Food Eng.* **2016**, *171*, 200–207. [CrossRef]

59. Garvín, A.; Ibarz, R.; Ibarz, A. Kinetic and thermodynamic compensation. A current and practical review for foods. *Food Res. Int.* **2017**, *96*, 132–153. [CrossRef]

60. Olivares-Tenorio, M.-L.; Verkerk, R.; van Boekel, M.A.J.S.; Dekker, M. Thermal stability of phytochemicals, HMF and antioxidant activity in cape gooseberry (*Physalis peruviana* L.). *J. Funct. Foods* **2017**, *32*, 46–57. [CrossRef]

 *antioxidants*

Article

Comparison of the Total Polyphenol Content and Antioxidant Activity of Chocolate Obtained from Roasted and Unroasted Cocoa Beans from Different Regions of the World

Bogumiła Urbańska * and Jolanta Kowalska

Faculty of Food Sciences, Department of Biotechnology, Microbiology and Food Evaluation,
Warsaw University of Life Sciences, 159 Nowoursynowska str., 02-787 Warsaw, Poland
* Correspondence: bogumila_urbanska@sggw.pl

Received: 1 July 2019; Accepted: 31 July 2019; Published: 6 August 2019

Abstract: The polyphenol content of cocoa beans and the products derived from them, depend on the regions in which they are grown and the processes to which they are subjected, especially temperature. The aim of the study was to compare the total content of polyphenols and antioxidant activity of chocolates obtained from roasted and unroasted cocoa beans. The chocolates produced from each of the six types of unroasted beans and each of the five types of roasted beans were investigated. The seeds came from Ghana, Venezuela, the Dominican Republic, Colombia and Ecuador. The highest total polyphenol content was determined in cocoa beans originating from Colombia and in the chocolates obtained from them. A higher content of total polyphenols was found in unroasted cocoa beans, which indicates the influence this process had on the studied size. The ability to scavenge free DPPH radicals was at a high level in both the beans and the chocolates produced from them, irrespective of the region where the raw material was grown. A positive correlation between the total polyphenol content and the ability to scavenge free radicals was found.

Keywords: cocoa; chocolate; polyphenols; antioxidants

---

1. Introduction

Chocolate is one of the most valued food products in the world [1]. Bitter chocolate in liquid form was already discovered in South America over 3000 years ago. This valuable product is made from cocoa liquor and fat (cocoa butter), with the addition of sugar, as well as milk and other additives (depending on the type of chocolate).

The main ingredient of chocolate is cocoa liquor, which is a mixture of fat and non-fat ingredients from the processing of cocoa beans. Furthermore, chocolate is produced with the addition of cocoa powder, which was recognised by EFSA as one of the richest sources of polyphenols [2]. Therefore, cocoa has recently become the target of increased scientific research, due to its pro-health properties.

Fresh cocoa beans contain about 32–39% water, 30–32% fat, 10–15% protein, 5–6% polyphenols, 4–6% pentosans, 2–3% cellulose, 2–3% sucrose, 1–2% theobromine, 1% acids and less than 1% caffeine. It is also a rich source of mineral components [3,4].

Three types of flavonoids dominate in cocoa beans: Proanthocyanins (circa 58%), catechins or flavan-3-ols (circa 37%) and anthocyanins (circa 4%) [5,6].

The high polyphenol content of cocoa, combined with its wide presence in many food products, makes it particularly interesting from a nutritional and health point of view [6–11].

The content and composition of polyphenols differs depending on the genotype, origin, growth conditions, degree of ripeness of the cocoa fruit and the grain processing parameters [12–15]. Knowledge of the changes in the polyphenol content throughout the technological chain, enables us to take steps to maintain the highest possible polyphenolic content in the final product.

Polyphenols shape not only antioxidant properties, but also affect sensory properties such as colour and taste [14,16,17]. During all stages of processing, the polyphenols present in cocoa beans may undergo many transformations, including polymerization, hydrolysis or reactions with proteins [18].

Roasting is a technological process which, due to its high temperature, is one of the most important in shaping the quality and sensory properties of cocoa beans and the products derived from them. During roasting, the structure of the beans is dried and loosened to remove the husks. Grinding of the kernels and pumping of the cocoa fat follows [19].

The roasting time lasts from 5 to 120 min (usually 10 to 35 min), at temperatures ranging from 110 °C to 160 °C (usually 120 °C–140 °C) [7,14,20]. High temperatures and the dehydration of cocoa beans in the roasting process, reduce the concentration of polyphenols and many volatile acids, especially acetic acid, which is responsible for the acidity of the product, as well as the bitter and astringent taste [21,22].

The roasting process also stimulates protein degradation, the synthesis of sulfur compounds, the Maillard reaction and the caramelization of sugars. These reactions allow new compounds to form, which contributes to the characteristic aroma, taste and colour of chocolate [23,24]. Oracz, et al. [25] stated that during the roasting process, conditions such as time and temperature influence phenolic stability, as well as the characteristics of the obtained taste.

Cocoa beans that are only fermented and dried (not roasted, referred to as "raw") during the production process contain many more phenolic compounds, which can have a positive effect on human health [18]. Omitting the roasting process, and not using high temperatures during the production of raw chocolates, results in the inability to evaporate many volatile components from the grain, including acetic acid, as well as increasing the acidity of such grains. The production of this type of chocolate does not use alkalization either, so the process of conching is often extended to up to four days, which makes it possible to obtain a final product that has a structure similar to traditional chocolates [26,27].

Roasting whole beans requires more energy than roasting smaller pieces, and at the same time makes it difficult to roast the beans evenly. Differences in roasting greatly influence the final product; i.e., the properties of chocolate, as well as its sensory and physicochemical properties. That is why manufacturers often roast a shot; i.e., ground beans. In this study, the antioxidant properties of beans that were roasted whole and ground, originating from the same region, were compared.

A number of studies are currently underway to develop a technology for roasting cocoa beans and cocoa middlings that will result in a semi-finished product with the least possible reduction in bioactive compounds and the lowest possible content of anti-nutritional compounds, such as acrylamide and acrolein [28].

The aim of this study was to compare the total polyphenol content and its ability to inactivate free DPPH radicals in chocolates produced from roasted and unroasted cocoa beans from different regions.

2. Methodology

2.1. Experimental Material

The experimental material consisted of roasted and unroasted cocoa beans, as well as the bitter chocolate obtained from the beans, from the Dominican Republic, Ghana, Venezuela, Colombia and Ecuador. Presented in Table 1 is a list of the research material and its labelling. The beans came from the 2014 harvest and were submitted for testing by one of the chocolate producers in Poland.

Table 1. List and labelling of test material No.

No	Abbreviation	Full Name
1	E-r	Ecuador-roasted beans
2	E-ur	Ecuador-unroasted beans
3	E-c	Ecuador-roasted and crushed beanss
4	E-ch	Ecuador-chocolate-roasted beans
5	C-r	Colombia-roasted beans
6	C-ur	Colombia-unroasted beans
7	C-c	Colombia-roasted and crushed beanss
8	C-ch	Colombia-chocolate-roasted beans
9	G-r	Ghana-roasted beans
10	G-ur	Ghana-unroasted beans
11	G-c	Ghana-roasted and crushed beanss
12	G-ch	Ghana-chocolate-roasted beans
13	D-r	Dominican-roasted beans
14	D-ur	Dominican-unroasted beans
15	D-c	Dominican-roasted and crushed beanss
16	D-ch	Dominican-chocolate-roasted beans
17	V-r	Venezuela-roasted beans
18	V-ur	Venezuela-unroasted beans
19	V-c	Venezuela-roasted and crushed beanss
20	V-ch	Venezuela-chocolate-roasted beans
21	P-ur	Peru-unroasted beans
22	P-ch	Peru-chocolate-unroasted beans

In order to illustrate the appearance of roasted and unroasted beans from different regions of the world, photographs were taken (shown in Figure 1).

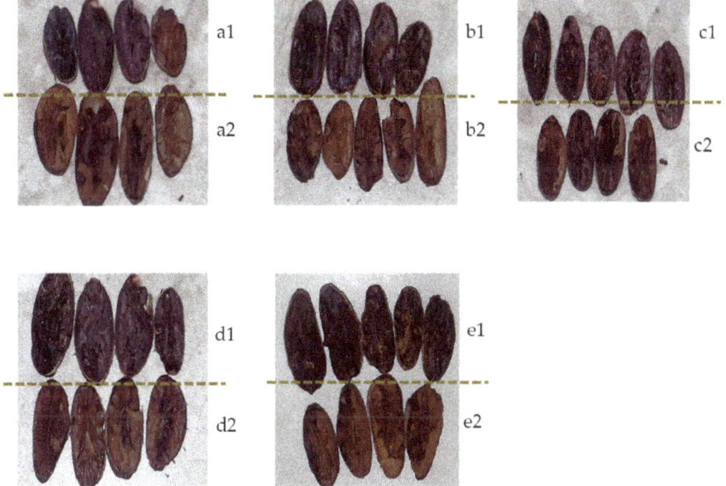

Figure 1. Cocoa bean cross-section photos (taken by authors); (**a1**) unroasted Colombia, (**a2**) roasted Colombia, (**b1**) unroasted Dominican Republic, (**b2**) roasted Dominican Republic, (**c1**) unroasted Ecuador, (**c2**) roasted Ecuador, (**d1**) unroasted Ghana, (**d2**) roasted Ghana, (**e1**) unroasted Venezuela, (**e2**) roasted Venezuela.

2.2. Analytical Methods

Chemical analyses have been carried out in at least three parallel repetitions. Total polyphenol content was determined by Folin-Ciocialteu's method [28]. Based on preliminary tests, a 70% acetone solution was used as a solvent to prepare extracts.

The extracts were prepared by weighing about 5 g of crushed test material into 300 mL grinding conical flasks and adding 100 mL of 70% acetone (*v/v*).

The samples were then shaken for 30 min in a Multi-Shaker PSU 20 Biosan shaker. Following this procedure, the solutions were filtered through the corrugated filters into 100 mL grinding flasks. In order to determine the total polyphenol content, 300 μL of the extract was taken from the tubes, and 4.15 mL of deionized water, 500 μL of 20% sodium carbonate solution and 50 μL of Folina-Ciocialteu reagent were added.

The blank sample was prepared by sampling: 300 μL extraction solution, 4.15 mL deionized water, 500 μL 20% sodium carbonate solution and 50 μL Folina-Ciocialteu reagent. Absorbance was measured at 700 nm on a SHIMADZ UV-1201V spectrophotometer. The apparatus was zeroed to a blank.

In order to calculate the total polyphenol content, a standard curve was prepared. Based on the results obtained, the graphical dependence of the absorbance of the solution on the amount of gallic acid contained in it was plotted. The total polyphenol content was calculated on the basis of the calibration curve and expressed in gallic acid equivalent per 100 g of product.

Determinating the ability of extracts to inactivate stable DPPH radicals

The extracts were prepared by weighing them into 300 mL grinding conical flasks of about 5 g of crushed test material and adding 100 mL of 70% acetone (*v/v*). The samples were then shaken for 30 min in a Multi-Shaker PSU 20 Biosan shaker. Following that, the solutions were filtered through the corrugated filters into 100 mL grinding flasks. Acetone extract (4 mL) and DPPH solution (1 mL), were taken to determine the appropriate sample. Acetone extract (4 mL) and methanol (1 mL), were collected for the blank sample. The samples were mixed and left to stand for 30 min, then absorbance was measured on the NOVASPEC II Pharmacia spectrophotometer (zeroing the apparatus for the blank test) at a wavelength of 562 nm in glass cuvettes with a diameter of 1 cm [29].

The antioxidant activity of the extracts against DPPH was calculated using the formula:

$$\text{Act.} = [(Ak - Awł)/Ak] \times 100\%. \tag{1}$$

where Act.-ntioxidant activity (%); Ak—the absorbance of the control sample; and Awł—absorption of the specific sample.

Microsoft Excel 2013 for Windows 10 was used to calculate the average values of the obtained results and to create graphs. Statistical analysis of the obtained results, correlation tests and the significance of the differences between the test samples were carried out using Statistica 13.0 using Tukey's test at the level of significance $p < 0.05$.

3. Results and Discussion

3.1. Determination of Total Polyphenols in Cocoa Beans

A total content of polyphenols in the research material was determined. The results were calculated in terms of gallic acid equivalent, as shown in Figure 2. The results for roasted cocoa beans in their entirety (marked with the letter "r"), middlings-crushed cocoa beans (marked with the letter "c") and whole unroasted cocoa beans (marked with the letters "ur") were presented.

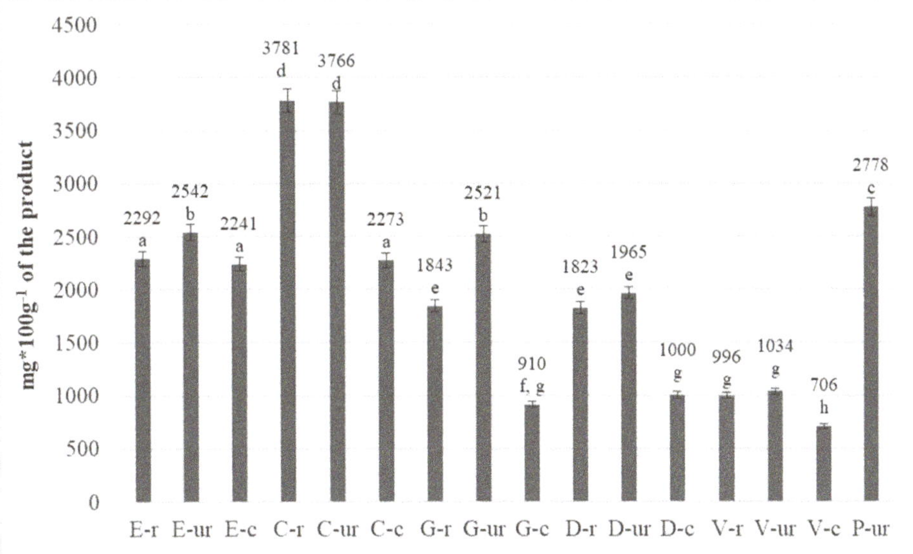

Figure 2. Total polyphenolic content in roasted and unroasted cocoa beans from different regions of the world (the same letter means no statistically significant differences between the analysed products at the level of significance α = 0.05; abbreviations used in the graph are described in Table 1).

The highest polyphenol content was found in both roasted and unroasted cocoa beans originating from Colombia, respectively 3781 mg/100 g of product and 3766 mg/100 g of product. The lowest total polyphenol content was found in roasted and unroasted cocoa beans originating from Venezuela (996 and 1034 mg/100 g of product). These differences can be explained by factors such as plant variety, geographical region, degree of maturity and post-harvest conditions [24]. The study also included cocoa beans that were organically farmed from Peru; ones which are used only in unroasted forms to produce raw chocolates. The polyphenol content of these beans was determined to be 2778 mg, the second highest after that of Colombia. The polyphenol content of roasted cocoa meal was also analysed, and the results obtained for whole roasted beans were compared. In the majority of analysed samples, the polyphenol content in crushed beans was significantly lower. The differences could result from a greater loss of these compounds due to the influence of oxygen and light on a smaller surface area. Additionally, the interior of the grain was discovered, to some extent, as a result of grinding, which additionally facilitated the influence of external factors on phenolic compounds (Figure 2).

In a study by Salvador, et al. [30], cocoa beans were analysed during production at one of the production plants in Brazil. In raw beans, over 6000 mg/100 g of product was determined, while in roasted beans from the same production cycle only about 1050 mg/ 100 g of product was determined.

In all types of unroasted grains, except for Colombia's (a similar level was determined), a higher content of polyphenols in grain was found, compared to the amount in roasted grains. The biggest difference was discovered in the beans from Ghana, and was almost 30% higher before roasting, whereas in Venezuela the difference was about 4% in favour of unroasted beans.

There are many known cases in literature where these differences are much greater. According to Medeiros, et al. [8], as a result of technological parameters and operations in the final product, the loss of flavonoids derived from cocoa beans can reach up to 80% of their initial content.

According to Gültekin-Özgüven, et al. [31] roasting and alkalization of cocoa beans reduces polyphenolic content by 65% and 87% respectively.

The determined polyphenol content may also be affected by the type of extraction used, the length of the procedure, the solvent used and the degree of fragmentation of the research material [32,33].

Statistical analysis showed significant differences in the content of polyphenols with a confidence level of 95%. Additionally, through the analysis of variance (ANOVA) for the one-way experiment, it was found that the type of cocoa beans (p-Value < 0.05) had a significant effect on the content of polyphenols. The beans from Colombia and Peru, which are separate homogeneous groups and have the highest polyphenolic compound content, deserve special attention. Based on the Tukey HDS test, eight homogeneous groups, marked with the same letters in Figure 2, were distinguished from cocoa beans.

The regression analysis showed a positive correlation (0.36) between the type of cocoa beans used in production and the content of polyphenols in chocolate.

3.2. Determination of Total Polyphenols in Chocolates

The results of total polyphenolic content in chocolates converted to gallic acid equivalent are presented in Figure 3. The polyphenolic content in chocolates ranged from 910 mg/100 g of chocolate product produced from roasted beans from Venezuela to 4055 mg/100 g of chocolate product from roasted beans from Columbia. The most chocolates had a higher polyphenol content than the amounts indicated in the cocoa beans. The process of chocolate production is complex and depends on many factors. The main source of phenolic compounds is the raw material-cocoa beans. According to literature data, most polyphenols undergo degradation during high-temperature processes. It should be remembered, however, that some polyphenols, e.g, (−)-epicatechin, may form complex, insoluble complexes, which are very difficult to determine analytically. At the same time, due to high temperatures, complex procyanidins may degrade to monomers, which in turn are determined analytically, thus affecting the overall polyphenol content of the product. In addition, bitter chocolates are often accompanied by the addition of a skimmed cocoa powder, which is an excellent source of polyphenols, and their positive health effects have been confirmed in the EFSA report [2].

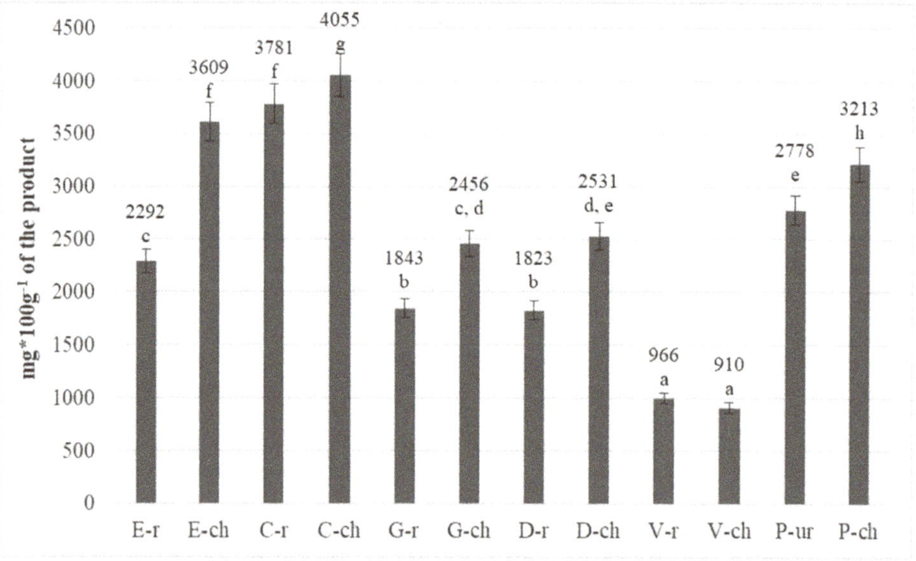

Figure 3. Comparison of the total polyphenol content in beans from different regions of the world and in the chocolates obtained from them (the same letter means that there are no statistically significant differences between the analysed products at a confidence level of $\alpha = 0.05$; the abbreviations used in the graph are described in Table 1).

Cocoa crops in Colombia currently have a high genetic variability, due to the crosses that can be found in some cocoa varieties (Stranger/Amazonian and Trinitario clones) [34]. As a result, the ecoclimatic conditions in the territory of Colombia are conducive to crop expansion into the country, and the increased capability to produce cocoa ecotypes with different bioactive profiles and flavours that are classified as high quality beans [34].

Polyphenols are contained in the non-fat components of cocoa beans, so it should be remembered that increasing the proportion of cocoa liquor results in an increase in the content of polyphenols in chocolates [35]. Jabłońska-Ryś, et al. [35] determined 2241–2746 mg of polyphenols per 100 g of product in dessert chocolate containing 70% cocoa, while in bitter choclates containing 75–80% cocoa, 2164–3129 mg of polyphenols were found per 100g of product.

In the present study, only chocolate from Venezuelan beans (910 mg/100 g product) showed lower results than those reported in the literature. In other cases, values higher than those presented in the studies by Jabłońska-Ryś, et al [35] or Kowalska and Sidorczuk [36] were obtained. The studies conducted by Meng, et al. [37] confirmed the influence of cocoa liquor's mass in chocolate on the content of polyphenols (578 mg/100 g in dark chocolates, while in milk and white chocolates the amount was 160 and 126 mg/100 g respectively). Studies by Cooper, et al. [38] also showed that apart from one chocolate from Venezuelan beans, the content of polyphenols was lower than in this study. On the other hand, Żyżelewicz, et al. [18] showed that dark chocolates produced with and without the addition of cocoa mass, prepared from unroasted cocoa beans, contained a higher concentration of total polyphenols (360 mg/100 g) compared to chocolate produced on the basis of roasted beans, in which only 841 mg of polyphenols per 100 g of the product was determined. The authors claim that during the preparation of chocolate, polyphenols may undergo many transformations, including polymerization and hydrolysis, as well as interacting with proteins and the products of the Maillard reaction.

A study by Lucia Godočiková, et al. [39] noted that chocolate produced in the traditional way (the roasting stage included) had almost twice the polyphenol content compared to cold processed products. Polyphenols can break down (degrade) or condense into complex compounds as a result of high temperatures. The lack of a roasting stage, for which the temperature of 110–160 °C is applied, may reduce the content of polyphenolic compounds, which was confirmed by Todorovic, et al. [40] in their studies. The tests were carried out for chocolates containing 65 to 75% cocoa. In the Jalil and Ismail [41] studies, chocolates obtained from unroasted beans were characterized by indirect polyphenol content in comparison to other analysed products. The amount of the analysed component in chocolates confirmed that the content of bioactive compounds depends on the technological process, raw material composition of the final product and the origin of cocoa beans [41].

The amount of polyphenols determined could also be affected by the type of extraction used, extraction time, the solvent used and the degree of fragmentation of the research material. Benayad, et al. [42], Cheng et al. [43] and Boulekbache-Makhlouf, et al. [44] showed that the use of acetone, in comparison to other polar organic compounds, increased the extraction of flavonoids and flavonoids from different plant materials.

The results of many scientific studies show how complex the process of determining the polyphenols content in chocolates is. Therefore, it is not clear which cocoa beans are the best source of polyphenols.

The statistical analysis of chocolates in the range of polyphenol content showed that only in one case (products from Venezuela), there was no statistically significant difference between the size determined in chocolates and beans (Figure 3). The remaining chocolates differed significantly from each other statistically with a confidence level of 95%. The analysis of variance (ANOVA) for the one-way experiment showed that the region of origin of cocoa beans significantly influenced the content of polyphenolic compounds in chocolates obtained from them (p-value < 0.05). Based on the Tukey HDS test, eight homogeneous groups, marked with the same letters on the data labels, were distinguished among chocolates.

Based on the results obtained in the present study and literature data, it can be concluded that the content of polyphenols depends on many factors, both those resulting from the genotype and region of raw material cultivation, as well as the technological processes and parameters used. Despite cultivation of the same varieties under similar conditions and the application of similar processing conditions, the content of polyphenols varies. It should be remembered that plants produce polyphenols in response to stresses, which can be very strong sunlight, drought, pest infestation and many others. In addition, the polyphenol content varies depending on the period of harvest. Cocoa beans from different plantations are mixed, and further processed as such. Therefore, there are differences in the polyphenols content of both the raw materials and the products derived from them. The analysis of the polyphenol content in the beans from different regions, taking into account the applied technological processes and their parameters, may be useful in creating blends and optimising the quality characteristics of the chocolates obtained from them.

3.3. Determination of the Ability of Extracts to Inactivate Stable DPPH Radicals in Cocoa Beans and Chocolates Derived Therefrom

The main polyphenols in cocoa beans are catechins, epicatechins, anthocyanins and procyanidins, the presence of which affets antioxidant activity [28].

Figures 4 and 5 show the results of the analysis of the activity of the tested extracts against DPPH radicals. The results of antiradical activity of antioxidant compounds in relation to DPPH radicals were calculated on the basis of measured absorbance values.

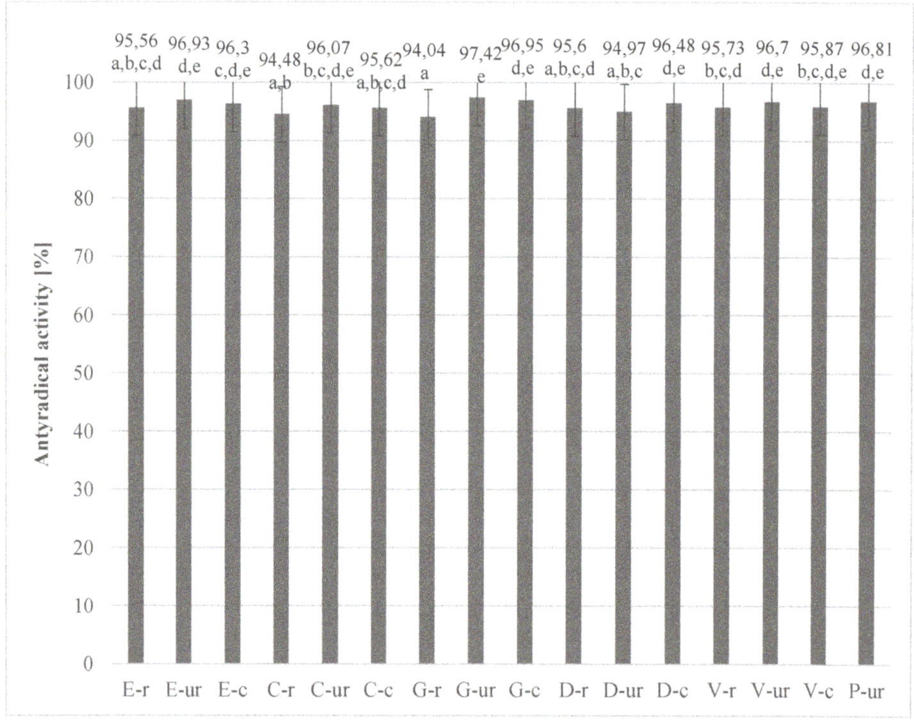

Figure 4. Anti-radical activity of grain extracts from different regions of the world (the same letter means that there are no statistically significant differences between the analysed products at a confidence level of $\alpha = 0.05$; the abbreviations used in the graph are described in Table 1).

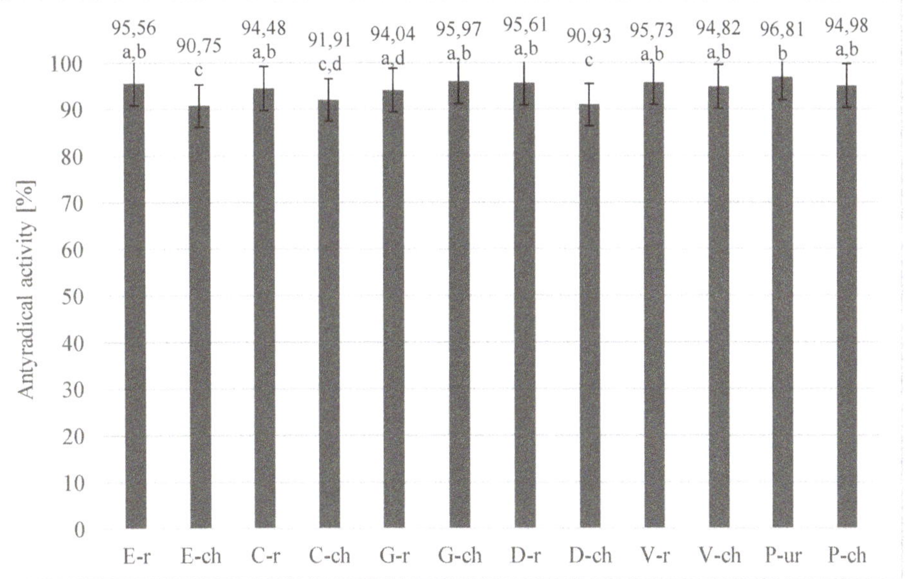

Figure 5. Comparison of anti-radical activity of grain extracts from different regions of the world and of the chocolate extracts produced from them (the same letter means that there are no statistically significant differences between the analysed products at a confidence level of $\alpha = 0.05$; the abbreviations used in the graph are described in Table 1).

Both the beans (roasted and unroasted) and chocolates produced from them were characterized by high ability to scavenge stable DPPH radicals. The analysis showed the influence of the roasting process on the antioxidant activity. In all roasted cocoa beans the ability to extinguish stable DPPH radicals was lower. Moreover, contrary to the determination of polyphenol content, higher antioxidant activity was determined in roasted meal in most of the analyzed samples than in whole grains. Referring to the results obtained for raw chocolate from Peru beans, it should be noted that despite the high content of polyphenols, the ability to extinguish the stable DPPH radicals was indirect. This is further evidence that the antioxidant properties of chocolate are an extremely complex phenomenon, dependent on a number of factors, and therefore it is not possible to determine the polyphenolic content and antioxidant activity unequivocally. Despite the separation of homogeneous groups, no statistically significant differences in antioxidant activity against stable DPPH radicals were found for cocoa beans studied, which was confirmed by statistical analysis (p-value = 0.148). The lowest antioxidant activity to extinguish DPPH radicals was obtained for chocolates from Ecuador (90.75%), Colombia (91.91%) and Dominican Republic (90.93%). Particularly noteworthy is chocolate from Colombia, which was characterized by the highest content of polyphenols in total. These compounds, according to literature data and the results of the correlation carried out in this study, significantly shape the antioxidant activity.

The above data show that traditional roasting significantly reduces both the concentration of polyphenols and the antioxidant activity of cocoa beans. These observations are consistent with the reports of Djikeng, et al. [45] who demonstrated that the decrease in antioxidant activity of cocoa beans during roasting was associated with the destruction of polyphenols contained in them. Roasting is considered one of the stages in the processing of cocoa beans that leads to a high loss of phenolic compounds and a decrease in antioxidant activity, as confirmed by Bauerin, et al. [46]. Similar conclusions were reached by Arlorio, et al. [47], who compared the antioxidant capacity of roasted and unroasted cocoa beans. Hu, et al [48] reported a decrease in antioxidant activity of between 44 and 50%

from high-temperature roasting (190 °C). According to Gültekin-Özgüven, et al. [32], roasting cocoa beans reduced the antioxidant capacity of DPPH by 21% and of ORAC by 51%. According to the results of the Kowalska and Sidorczuk [38] studies, the tested cocoa beans and chocolates were characterized by a high ability to scavenge free radical DPPH at the level of 88–92%. These results are slightly lower than the results obtained in this study (cocoa beans above 95%, and chocolate from 90% to 96%), which may be due to the different content of cocoa components and the region of origin of the beans. Żyżelewicz et al. [49], in their studies, showed that the antioxidant activity of chocolate decreased with an increase in the percentage of prepared cocoa liquor from unroasted beans. Wollgast, et al. [50] found that the evaluation of polyphenolic compounds and antioxidant activity depends largely on the solvent and extraction procedure, which is not standardized in the cocoa literature, so the data are difficult to compare. According to Di Mattia, et al. [29], discrepancies in total phenol (TPC) colorimetric tests may occur due to phenolic compounds being used as reference for the standard curve and the presence of reducing compounds that interfere with the test. Therefore, comparison of antioxidant activity results may be problematic due to the large number of heterogeneous tests used.

The regression analysis at a 95% confidence interval showed that there is a positive correlation (0,86) between the content of polyphenols in chocolate and the ability to inactivate stable DPPH radicals. Additionally, a very strong correlation ($r = 0.72$) between the antiradical activity of grain extracts from different regions of the world and chocolate extracts produced from them was demonstrated on the basis of a statistical analysis.

4. Summary and Conclusions

Based on the research conducted and results obtained, it is now known that roasted cocoa beans in most of the analyzed samples had a lower polyphenol content than unroasted grains. However, the content of polyphenols in chocolates was much higher than in the cocoa beans from which they were obtained. The ability to extinguish free DPPH radicals was at a high level both in the beans and chocolates, and was higher than in the literature's data. This ability decreased after roasting the cocoa beans, and after the whole process of chocolate production, in relation to the beans from which it was produced.

The conducted research showcases the influence of the type of cocoa beans and the technological processes used on the properties of the chocolates obtained. Further analysis of chocolate products based on unroasted beans and the evaluation of their usefulness, in terms of dietary inclusion of products with specific antioxidant properties, seems justified.

Author Contributions: B.U. conceived the idea, compiled the information and drafted the manuscript, edited the article for submission; J.K. designed the study, provided guidance and revised the article for submission. All authors contributed in revision of manuscript. All authors read and approved the final manuscript.

Funding: This research did not receive any specific grant from funding agencies in the public, commercial, or not-for-profit sectors.

Conflicts of Interest: The authors declare no conflict of interest.

References

1. Caligiani, A.; Marseglia, A.; Prandi, B.; Palla, G.; Sforza, S. Influence of fermentation level and geographical origin on cocoa bean oligopeptide pattern. *Food Chem.* **2016**, *211*, 431–439. [CrossRef] [PubMed]
2. EFSA. Scientific Opinion on the substantiation of a health claim related to cocoa flavanols and maintenance of normal endothelium-dependent vasodilation pursuant to Article 13(5) of Regulation (EC) No 1924/2006. *EFSA J.* **2012**, *10*, 2809–2830. [CrossRef]
3. Bertazzo, A.; Agnolin, F.; Comai, S.; Zancato, M.; Costa, C.V.; Seraglia, R.; Traldi, P. The protein profile of *Theobroma cacao* L. seeds as obtained by matrix-assisted laser desorption/ionization mass spectrometry. *Rapid Commun. Mass Spectrom.* **2011**, *25*, 2035–2042. [CrossRef] [PubMed]
4. Kruszewski, B.; Obiedziński, M.W. Multivariate analysis of essential elements in raw cocoa and processed chocolate mass materials from three different manufacturers. *LWT* **2018**, *98*, 113–123. [CrossRef]

5. Khan, N.; Nicod, N.M. Biomarkers of cocoa consumption. In *Chocolate and Health*; Springer: Milano, Italy, 2012; pp. 33–40.

6. Khan, N.; Khymenets, O.; Urpí-Sardà, M.; Tulipani, S.; Garcia-Aloy, M.; Monagas, M.; Mora-Cubillos, X.; Llorach, R.; Andres-Lacueva, C. Cocoa polyphenols and inflammatory markers of cardiovascular disease. *Nutrients* **2014**, *6*, 844–880. [CrossRef] [PubMed]

7. Ioannone, F.; Di Mattia, C.D.; De Gregorio, M.; Sergi, M.; Serafini, M.; Sacchetti, G. Flavanols, proanthocyanidins and antioxidant activity changes during cocoa (*Theobroma cacao* L.) roasting as affected by temperature and time of processing. *Food Chem.* **2015**, *174*, 256–262. [CrossRef] [PubMed]

8. Da Silva Medeiros, N.; Marder, R.K.; Farias Wohlenberg, M.; Funchal, C.; Dani, C. Total Phenolic Content and Antioxidant Activity of Different Types of Chocolate, Milk, Semisweet, Dark, and Soy, in Cerebral Cortex, Hippocampus, and Cerebellum of Wistar Rats. *Biochem. Res. Int.* **2015**, 2015. [CrossRef] [PubMed]

9. Cinquanta, L.; Di Cesare, C.; Manoni, R.; Piano, A.; Roberti, P.; Salvatori, G. Mineral essential elements for nutrition in different chocolate products. *Int. J. Food Sci. Nutr.* **2016**, *67*, 773–778. [CrossRef]

10. Giacometti, J.; Muhvić, D.; Pavletić, A.; Dudarić, L. Cocoa polyphenols exhibit antioxidant, anti-inflammatory, anticancerogenic, and anti-necrotic activity in carbon tetrachloride-intoxicated mice. *J. Funct. Foods* **2016**, *23*, 177–187. [CrossRef]

11. Tsang, C.; Hodgson, L.; Bussu, A.; Farhat, G.; Al-Dujaili, E. Effect of Polyphenol-Rich Dark Chocolate on Salivary Cortisol and Mood in Adults. *Antioxidants* **2019**, *8*, 149. [CrossRef]

12. Rusconi, M.; Conti, A. *Theobroma cacao* L., the Food of the Gods: A scientific approach beyond myths and claims. *Pharmacol. Res.* **2010**, *61*, 5–13. [CrossRef] [PubMed]

13. Katz, D.L.; Doughty, K.; Ali, A. Cocoa and Chocolate in Human Health and Disease. *Antioxid. Redox Sign.* **2011**, *15*, 2779–2811. [CrossRef] [PubMed]

14. Kothe, L.; Zimmermann, B.F.; Galensa, R. Temperature influences epimerization and composition of flavanol monomers, dimers and trimers during cocoa bean roasting. *Food Chem.* **2013**, *141*, 3656–3663. [CrossRef] [PubMed]

15. Żyżelewicz, D.; Krysiak, W.; Oracz, J.; Sosnowska, D.; Budryn, G.; Nebesny, E. The influence of the roasting process conditions on the polyphenol content in cocoa beans, nibs and chocolates. *Food Res. Int.* **2016**, *89*, 918–929. [CrossRef]

16. Dabas, D. Polyphenols as colorants. *Adv. Food Technol. Nutr. Sci. Open J.* **2016**, *SE*, S1–S6. [CrossRef]

17. Giacometti, J.; Jolić, S.M.; Josić, D. Cocoa Processing and Impact on Composition. In *Processing and Impact on Active Components in Food*; Elsevier Inc.: Amsterdam, Netherlands, 2014; pp. 605–612. [CrossRef]

18. Zyzelewicz, D.; Budryn, G.; Oracz, J.; Antolak, H.; Kregiel, D.; Kaczmarska, M. The effect on bioactive components and characteristics of chocolate by functionalization with raw cocoa beans. *Food Res. Int.* **2018**, *113*, 234–244. [CrossRef] [PubMed]

19. Afoakwa, E.O.; Quao, J.; Takrama, J.; Budu, A.S.; Saalia, F.K. Chemical composition and physical quality characteristics of Ghanaian cocoa beans as affected by pulp pre-conditioning and fermentation. *J. Food Sci. Technol.* **2013**, *50*, 1097–1105. [CrossRef] [PubMed]

20. Farah, D.M.H.; Zaibunnisa, A.H.; Misnawi, J.; Zainal, S. Effect of roasting process on the concentration of acrylamide and pyrazines in roasted cocoa beans from different origins. *APCBEE Procedia* **2012**, *4*, 204–208. [CrossRef]

21. Ramli, N.; Hassan, O.; Said, M.; Samsudin, W.; Idris, N.A. Influence of roasting conditions on volatile flavor ofroasted Malaysian cocoa beans. *J. Food Process. Preserv.* **2006**, *30*, 280–298. [CrossRef]

22. Cadena Cala, T.; Herrera Ardila, Y.M. *Evaluación del Efecto del Procesamiento del Cacao Sobre el Contenido de Polifenoles y su Actividad Antioxidant*; Trabajo de Grado, Escuela de Química, Universidad Industrial de Santander: Bucaramanga, Colombia, 2008.

23. Hoskin, J.C.; Dimick, P.S. Role ofnonenzymatic browning during the processing ofchocolate—A review. *Process Biochem.* **1984**, *19*, 92–104.

24. Oracz, J.; Nebesny, E. Effect of roasting parameters on the physicochemical characteristics of high-molecularweight Maillard reaction products isolated from cocoa beans of different *Theobroma cacao* L. groups. *Eur. Food Res. Technol.* **2019**, *245*, 111–128. [CrossRef]

25. Oracz, J.; Nebesny, E. Influence of roasting conditions on the biogenic amine content in cocoa beans of different *Theobroma cacao* cultivars. *Food Res. Int.* **2014**, *55*, 1–10. [CrossRef]

26. Albak, F.; Tekin, A.R. The effect of addition of ingredients on physical properties of dark chocolate during conching. *Basic Res. J. Food Sci. Technol.* **2014**, *1*, 51–59.

27. García-Alamilla, P.; Salgado-Cervantes, M.A.; Barel, M.; Berthomieu, G.; Rodríguez-Jimenes, G.C.; García-Alvarado, M.A. Moisture, acidity and temperature evolution during cacao drying. *J. Food Eng.* **2007**, *79*, 1159–1165. [CrossRef]

28. Di Mattia, C.D.; Sacchetti, G.; Mastrocola, D.; Serafini, M. From Cocoa to Chocolate: The impact of Processing on In Vitro Antioxidant Activity and the Effects of Chocolate on Antioxidant Markers in Vivo. *Front. Immunol.* **2017**, *8*, 1207. [CrossRef] [PubMed]

29. Singleton, V.L.; Rossi, J.A. Colorimetry of total phenolics with phosphomo- lybdic-phosphotungstic acid reagents. *Am. J. Enol. Viticult.* **1965**, *16*, 144–158.

30. Salvador, I.; Massarioli, A.P.; Silva, A.P.S.; Malaguetta, H.; Melo, P.S.; Alencar, S.M. Can we conserve trans-resveratrol content and antioxidant activity during industrial production of chocolate? *J. Sci. Food Agric.* **2019**, *99*, 83–89. [CrossRef]

31. Gültekin-Özgüven, M.; Berktaş, I.; Özçelik, B. Influence of processing conditions on procyanidin profiles and antioxidant capacity of chocolates: Optimization of dark chocolate manufacturing by response surface methodology. *LWT Food Sci. Technol.* **2016**, *66*, 252–259. [CrossRef]

32. Miller, K.B.; Stuart, D.A.; Smith, M.L.; Lee, C.Y.; McHale, N.L.; Flanagan, J.A.; Du, B.; Hurst, W.J. Antioxidant activity and polyphenol and procyanidin contents of selected commercially available cocoa containing and chocolate products in the United States. *J. Agric. Food Chem.* **2006**, *54*, 4062–4068. [CrossRef]

33. Payne, M.J.; Hurst, W.J.; Miller, K.B.; Rank, C.; Stuart, D.A. Impact of fermentation, drying, roasting, and Dutch processing on epicatechin and catechin content of cacao beans and cocoa ingredients. *J. Agric. Food Chem.* **2010**, *58*, 10518–10527. [CrossRef]

34. Carrillo, L.C.; Londoño-Londoño, J.; Gil, A. Comparison of polyphenol, methylxanthines and antioxidant activity in Theobroma cacao beans from different cocoa-growing areas in Colombia. *Food Res. Int.* **2010**, *60*, 273–280. [CrossRef]

35. Jabłońska-Ryś, E. Zawartość polifenoli w czekoladach. *Nauka Przyr. Technol.* **2012**, *6*, #30.

36. Kowalska, J.; Sidorczuk, A. Analysis of the effect of technological processing on changes in antioxidant properties of cocoa processed products. *Pol. J. Food Sci.* **2007**, *57*, 96–99.

37. Meng, C.C.; Jalil, A.M.M.; Ismail, A. Phenolic and theobromine contents of commercial dark, milk and white chocolates on the Malaysian market. *Molecules* **2009**, *14*, 200–209. [CrossRef] [PubMed]

38. Cooper, K.A.; Donovan, J.L.; Waterhouse, A.I.; Williamson, G. Cocoa and health: A decade of research. *Br. J. Nutr.* **2008**, *99*, 1–11. [CrossRef] [PubMed]

39. Godočiková, L.; Ivanišová, E.; Árvay, J.; Petrová, J.; Kačániová, M. The comparison of biological activity of chocolates made by different technological procedures. *Potravinarstvo* **2016**, *10*, 316–322. [CrossRef]

40. Todorovic, V.; Vidovic, B.; Todorovic, Z.; Sobajic, S. Correlation between Antimicrobial, Antioxidant Activity, and Polyphenols of Alkalized/Nonalkalized Cocoa Powders. *J. Food Sci.* **2017**, *82*, 1020–1027. [CrossRef]

41. Jalil, A.M.M.; Ismail, A. Polyphenols in Cocoa and Cocoa Products: Is There a Link between Antioxidant Properties and Health? *Molecules* **2008**, *13*, 2190–2219. [CrossRef]

42. Benayad, Z.; Gómez-Cordovés, C.; Es-Safi, N.E. Characterization of Flavonoid Glycosides from Fenugreek (Trigonella foenum-graecum) Crude Seeds by HPLC–DAD–ESI/MS Analysis. *Int. J. Mol. Sci.* **2014**, *15*, 20668–20685. [CrossRef]

43. Cheng, V.J.; Bekhit, A.E.-D.A.; McConnell, M.; Mros, S.; Zhao, J. Effect of extraction solvent, waste fraction and grape variety on the antimicrobial and antioxidant activities of extracts from wine residue from cool climate. *Food Chem.* **2012**, *134*, 474–482. [CrossRef]

44. Boulekbache-Makhlouf, L.; Slimani, S.; Madani, K. Total phenolic content, antioxidant and antibacterial activities of fruits of Eucalyptus globulus cultivated in Algeria. *Ind. Crops Prod.* **2013**, *41*, 85–89. [CrossRef]

45. Djikeng, F.T.; Teyomnou, W.T.; Tenyang, N.; Tiencheu, B.; Morfor, A.T.; Touko, B.A.H.; Houketchang, S.N.; Boungo, G.T.; Karuna, M.S.L.; Ngoufack, F.Z. Effect of traditional and oven roasting on the physicochemical properties of fermented cocoa beans. *Heliyon* **2018**, *4*, e00533. [CrossRef] [PubMed]

46. Bauer, D.; de Abreu, J.P.; Oliveira, H.S.S.; Goes-Neto, A.; Koblitz, M.G.B.; Teodoro, A.J. Antioxidant Activity and Cytotoxicity Effect of Cocoa Beans Subjected to Different Processing Conditions in Human Lung Carcinoma Cells. *Oxid. Med. Cell. Longev.* **2016**, *2016*, 1–11. [CrossRef] [PubMed]

47. Arlorio, M.; Locatelli, M.; Travaglia, F.; Coïsson, J.; Del Grosso, E.; Minassi, A.; Appendino, G.; Martelli, A. Roasting impact on the contents of clovamide (N-caffeoyl-L-DOPA) and the antioxidant activity of cocoa beans (*Theobroma cacao* L.). *Food Chem.* **2008**, *106*, 967–975. [CrossRef]
48. Hu, S.; Kim, B.Y.; Baik, M.Y. Physicochemical properties and antioxidant capacity of raw, roasted and puffed cacao beans. *Food Chem.* **2016**, *194*, 1089–1094. [CrossRef] [PubMed]
49. Zyzelewicz, D.; Bojczuk, M.; Budryn, G.; Jurgonski, A.; Zdunczyk, Z.; Juskiewicz, J.; Oracz, J. Influence of diet enriched with cocoa bean extracts on physiological indices of laboratory rats. *Molecules* **2019**, *24*, 825. [CrossRef] [PubMed]
50. Wollgast, J.; Anklam, E. Review on polyphenols in Theobroma cacao: Changes in composition during the manufacture of chocolate and methodology for identification and quantification. *Food Res. Int.* **2000**, *33*, 423–447. [CrossRef]

 *antioxidants*

Article

Study of Polyphenol Content and Antioxidant Properties of Various Mix of Chocolate Milk Masses with Different Protein Content

Bogumiła Urbańska [1,*], Tomasz Szafrański [1], Hanna Kowalska [2] and Jolanta Kowalska [1]

[1] Institute of Food Sciences, Department of Technology and Food Evaluation, Warsaw University of Life Sciences, 159c Nowoursynowska St., 02-776 Warsaw, Poland; jolanta_kowalska@sggw.pl (J.K.)
[2] Institute of Food Sciences, Department of Food Engineering and Process Management, Warsaw University of Life Sciences, 159c Nowoursynowska St., 02-776 Warsaw, Poland; hanna_kowalska@sggw.pl
* Correspondence: bogumila_urbanska@sggw.pl; Tel.: +48-608-16-19-50

Received: 11 March 2020; Accepted: 1 April 2020; Published: 3 April 2020

Abstract: The aim of the study was to analyze the antioxidant character of conched chocolate milk masses, taking into account different protein content in milk. For the study, cocoa liquor obtained from roasted and unroasted cocoa beans from different regions, as well as milk powder obtained by spray and cylindrical drying were used. The analysis that was carried out showed that the protein content of powdered milk products ranged from about 11.6% (*w/w*) to over 31% (*w/w*). Lower content of polyphenols and lower antioxidant activity were shown in the masses to which the addition of milk with higher protein content was applied. The analysis of antioxidant character of chocolate milk masses showed higher total polyphenols content in masses prepared from unroasted cocoa beans liquor.

Keywords: conching; milk chocolate; milk powder; protein; polyphenols; antioxidant activity

1. Introduction

Cocoa beans are seeds of the tropical Theobroma cacao tree. There are three best known varieties of this plant: Forastero, which covers about 95% of the world's cocoa production and is most commonly used to produce chocolate; Criollo, which is the most exclusive but at the same time the least cultivated variety due to its susceptibility to diseases; and Trinitario, a hybrid combining the characteristics of Criollo and Forastero, with an intense chocolate aroma, with a hint of wine, more resistant to diseases and pests [1,2].

Many studies and numerous publications confirm that cocoa beans are a raw material carrying a powerful load of antioxidants, valued in today's diets mainly for their antiatherogenic, antiradical, and anticancer properties [3–6]. The genotype of cocoa beans, the region in which they are grown, the environmental conditions, as well as the conditions and parameters of applied technological operations, especially fermentation, drying, roasting, and conching, have a significant influence on the formation of sensory characteristics and antioxidant properties of cocoa bean processing products [7–10]. As numerous studies have shown, the most degrading content of phenolic compounds is the roasting stage—high-temperature heating of beans or cocoa shots, which significantly affects the antioxidant potential of chocolates [11–15]. The use of less processed cocoa beans (omitting the roasting process) for the production of chocolate mass may result in a product with a higher content of polyphenols and higher antiradical activity, as well as positively change the nutritional and health value. Therefore, interest in cocoa mass obtained from unroasted cocoa beans [2] has increased in recent years.

Conching is one of the main and most important stages in the chocolate production process. It consists of stirring and aerating the chocolate mass with simultaneous heating at a certain temperature

(>40 °C) [16,17]. Conching plays an important role in the development of taste by removing undesirable volatile compounds and moisture and by obtaining a homogeneous mass of appropriate particle size [9]. The time of conching, which is unfavorable from the point of view of production efficiency, is significantly influenced by the temperature and speed of mixing [18]. These parameters are also important for the course and intensification of the Maillard reaction as well as the Strecker degradation reaction [19]. The choice of process parameters is adapted to the type of product, its composition, and the production capacity of the plant. In order to limit or prevent the Maillard reaction it is recommended to carry out the process of conching milk masses at a temperature not exceeding 50 °C [9]. The Maillard reaction is affected by many aspects, including temperature, pH, water content, duration of heating, type of reactant, oxygen, ratio of amino acid to sugar, metals, and reaction inhibitors [20].

Few studies indicate that the percentage of the conching process has no significant effect on the content and phenolic system, as well as on the antioxidant activity [1,19,21,22]. The results obtained by Di Mattia, et al. [23] even indicate an increase in the antioxidant activity of conched masses, mainly due to the growing potential of melanosides.

The legal act establishing guidelines and minimum requirements for a product that can be called chocolate is Directive 2000/36/EC of the European Parliament and of the Council of 23 June 2000 [24]. It indicates that chocolate is a product obtained from cocoa products and sugar, containing not less than 35% (*w/w*) total dry cocoa solids and a minimum of 18% (*w/w*) cocoa butter and 14% (*w/w*) dry non-fat cocoa solids. The original chocolate recipe is based on only three ingredients: cocoa mass, cocoa butter (fat), and sugar. However, with the development of the chocolate market, other raw materials and additives such as emulsifiers, which stabilize the structure and consistency of chocolate and flavors that enhance its taste and smell, started to be used [3,25]. An example of another raw material used in the production of chocolate is powdered milk, which was first used in 1875 and gave the chocolate a more velvety texture and a pleasant sweet and milky character. Two types of milk powder are used to produce chocolate: roll-dried and spray-dried. Roll-cured milk is more preferred because the chocolate mass produced from it ensures optimum viscosity, while spray-dried milk-based chocolate mass has a much higher viscosity. This is due, among other things, to physical differences in the material. Powder dried on a cylinder has a high content of free fat (at the level of about 90%), while spray-dried powder has below 10%. Powder dried on rolls usually has a larger average particle size of about 150 μm and a small vacuole volume, while spray-dried powder has a smaller particle size—about 70 μm—and a high vacuole volume [26].

The protein content in bitter chocolate is at the level of 5–15% (*w/w*) and it is the protein derived from cocoa beans. Peptides are currently considered important bioactive constituents of food but the potential biological activities of oligopeptides found in cocoa are not sufficiently researched in cocoa literature. This protein is characterized primarily by a low degree of amino acid release during digestion, which results in low nutritional importance of this component in our diet. In the case of milk chocolates, however, the protein content is primarily determined by the amount of milk powder added, in which the content of this component oscillates between 25% and 30% (*w/w*). Milk protein is characterized by much better assimilability compared to cocoa, which translates into increased nutritional importance of this protein in the diet of the consumer [27,28]. According to Maidannyk, et al. [29], milk powder proteins are susceptible to caking and adhesion and may be highly insoluble. Therefore, storage conditions of the powder are an important factor responsible for such problems.

Most studies analyze the antioxidant properties of bitter chocolates, which are a rich source of these compounds. Due to lower content of cocoa mass and fat-free components of cocoa bean processing, milk chocolates are characterized by much lower antioxidant potential [30]. Moreover, most of the studies concern chocolates obtained in the traditional technological cycle and include mainly the influence of the origin of raw materials and the roasting process on the physical and chemical properties of these products. The results of these studies are inconclusive and although they tend to confirm the positive correlation between polyphenol content and antioxidant activity, as well as the degrading effect of roasting on polyphenol content, there are studies that do not confirm these results [22]. As shown by the results of the work involving the evaluation of consumer preference for chocolate, consumers are most likely to reach for milk chocolates.

However, research on milk chocolates is not conducted on a large scale, especially in terms of evaluating their antioxidant potential. They are mainly produced using milk obtained from spray drying and such products are the subject of most studies and publications. Moreover, chocolates called "raw"—produced without roasting—have appeared on the market. There is little research into these products. As well as few studies on the effect of conching on antioxidant properties have been published. Therefore, it seems reasonable to attempt to assess the effect of conching on the antioxidant potential of chocolate milk masses, taking into account the protein content of milk powder obtained by spray or cylindrical drying, as well as to analyze the subject of research after using roasted and unroasted cocoa pulp.

The aim of the study was to investigate the relationship between the protein content of milk powder and cocoa mass and the antioxidant potential of chocolate milk masses after conching.

2. Materials and Methods

2.1. Test Material

The research material was 19 milk powders differing in production technologies (Table 1) and the raw materials used to prepare chocolate milk masses—cocoa liquors from 3 manufacturers, differing the country of origin and in the process of processing the beans from which they were made and selected powdered milk (Table 2).

Table 1. Dry matter and protein content in milk powder analyzed.

Milk Sample Number	Characteristics of Milk Type of Milk/Supplier/Milking Time/Production Technique	Protein Content (%)	Dry Matter Content (%)
1	WMP- MX/I 2018/D	28.53 ± 0.9	97.02 ± 0.08
2	WMP- MX/XII 2018/D	31.10 ± 1.1	97.35 ± 0.1
3	WMP- MX/IV 2017/D	30.35 ± 0.6	97.15 ± 0.2
4	WMP- MX/II 2018/D	27.44 ± 0.6	96.51 ± 0.1
5	WMP- MX/IX 2018/D	29.14 ± 0.8	96.72 ± 0.2
6	WMP- MX/VIII 2018/D	28.41 ± 1.0	96.34 ± 0.1
7	WMP- MX/V 2018/D	29.06 ± 0.8	97.31 ± 0.3
8	WMP- MX/IX 2018/D	29.00 ± 0.7	96.85 ± 0.09
9	WMP- MX/II 2018/D	27.99 ± 0.8	97.25 ± 0.1
10	WMP- MX/V 2018/D	28.86 ± 0.9	96.41 ± 0.2
11	WP- MX/III 2018/D	13.51 ± 1.1	97.91 ± 0.1
12	WMP- MX/I 2018/D	29.47 ± 0.9	95.92 ± 0.3
13	MP - MX/IV 2018/C (consisting in 80% of WM, enriched with lactose and WP)	18.90 ± 0.8	96.55 ± 0.09
14	MP - MX/IV 2018/C (consisting of 47% sugar and the remainder of milk, permeate and cream)	13.19 ± 1.0	97.26 ± 0.1
15	WMP- MX/VII 2018/D	25.78 ± 1.1	95.32 ± 0.2
16	WMP- MX/VI 2018/D	26.43 ± 0.9	95.17 ± 0.2
17	WMP- MX/V 2018/D	27.02 ± 0.6	96.24 ± 0.3
18	WMP- MZ/V 2018/D	26.76 ± 0.7	97.56 ± 0.09
19	WP- MZ/V 2018/D	11.61 ± 0.9	97.14 ± 0.1

The results of the protein content and dry weight given in grams in 100 g of product (g 100 g^{-1}) were expressed in (%)—(*w/w*). Whole-milk powder—WMP, Whey powder—WP, Milk preparation—MP, manufacturer X—MX, manufacturer Y—MY, manufacturer Z—MZ, drying technique—D, cylindrical drying technique—C.

Table 2. List and marking of raw materials used to prepare chocolate milk masses.

Sample Code	Characteristics of the Test Material
Ch_1	Cocoa liquor/unroasted beans/Peru,
Ch_2	Cocoa liquor/roasted beans/Ivory Coast
Ch_3	Cocoa liquor/roasted beans/Ghana
Ml_1	WMP- MX/IV 2017/D
Ml_2	WMP/- MX/II 2018/D
Ml_3	WMP/- MX/IX 2018/D
Ml_4	MP – MX/IV 2018/C (consisting in 80% of WM, enriched with lactose and WP)
Ml_5	MP – MX/IV 2018/C (consisting of 47% sugar and the remainder of milk, permeate and cream)
Ml_6	WMP- MX/VII 2018/D
Ml_7	WMP- MZ/V 2018/D

Whole-milk powder—WMP, Whey powder—WP, Milk preparation—MP, manufacturer X—MX, manufacturer Y—MY, manufacturer Z—MZ, drying technique—D, cylindrical drying technique—C.

Milk powders and cocoa masses were obtained from the chocolate products manufacturers.

The first stage of the research was to carry out the analysis of dry matter and protein content in powdered milk according to the research methodology described in the subchapter of analytical methods 3.1 and 3.2, respectively.

On the basis of the results obtained, seven milk products were selected for further studies—3, 4, 5, 13, 14, 15, and 18 differing in protein and dry matter content. The selected milk was marked with new codes respectively: 3—Ml_1, 4—Ml_2, 5—Ml_3, 13—Ml_4, 14—Ml_5, 15—Ml_6, and 18—Ml_7 (Table 2).

The samples marked with the codes Ml_1, Ml_2, Ml_3, Ml_6, and Ml_7 were spray-dried whole milk powders, differing in protein content (Table 1). On the other hand, the milk Ml_4 and Ml_5 differed from the other compositions (they were mixtures of milk with sugar) and were produced using the cylindrical method.

2.2. Technological Process

Chocolate milk masses were prepared by mixing (Table 3):

- Cocoa masses—16.2%,
- Cocoa butter—12.3%,
- Sugar—50%,
- Milk powder—18.0%,
- Whey—3.2%,
- Lecithins—0.3%.

Table 3. List and marking of chocolate milk masses.

Sample Code	Characteristics of Chocolate Milk Masses	Sample Code	Characteristics of Chocolate Milk Masses
T-1	CMM prepared using Ch_1 and Ml_1	T-12	CMM prepared using Ch_2 and Ml_5
T-2	CMM prepared using Ch_1 and Ml_2	T-13	CMM prepared using Ch_2 and Ml_6
T-3	CMM prepared using Ch_1 and Ml_3	T-14	CMM prepared using Ch_2 and Ml_7
T-4	CMM prepared using Ch_1 and Ml_4	T-15	CMM prepared using Ch_3 and Ml_1
T-5	CMM prepared using Ch_1 and Ml_5	T-16	CMM prepared using Ch_3 and Ml_2
T-6	CMM prepared using Ch_1 and Ml_6	T-17	CMM prepared using Ch_3 and Ml_3
T-7	CMM prepared using Ch_1 and Ml_7	T-18	CMM prepared using Ch_3 and Ml_4
T-8	CMM prepared using Ch_2 and Ml_1	T-19	CMM prepared using Ch_3 and Ml_5
T-9	CMM prepared using Ch_2 and Ml_2	T-20	CMM prepared using Ch_3 and Ml_6
T-10	CMM prepared using Ch_2 and Ml_3	T-21	CMM prepared using Ch_3 and Ml_7
T-11	CMM prepared using Ch_2 and Ml_4		

Chocolate milk masses—CMM, other code explanation in Table 2.

Chocolate milk masses were prepared from ingredients weighed and mixed at controlled temperature and time conditions in the Termomix device (Vorweck, Germany)—the household appliances. The first step was to liquefy a weighed portion of cocoa butter at 45 °C. Then the fat was poured into another container. The cocoa liquor was liquefied at 55 °C. The sugar, milk powder, whey, and 10% cocoa butter were added to the liquefied cocoa liquor. The ingredients were mixed for 10 min at a constant temperature of 45 °C, then residual cocoa fat and lecithin were added. The final stage was the proper conching, i.e., slow mixing of the mass at 55 °C for 1 h.

3. Analytical Methods

The reagents used for the analyses were purchased from Sigma Aldrich Company.

3.1. Determination of Dry Matter Content

The dry matter content was determined in all samples of milk powder and cocoa liquor. This analysis consisted in the evaporation of water from the material tested during the drying process, followed by weight determination of the residue (dry matter).

The weight of test material was dried to constant mass at 105 °C for 3 h in the laboratory chamber oven type SUP-65 WG manufacturer WAMED. Samples of cocoa liquor and chocolate masses were dried with sand to increase the evaporation surface. The dry matter content was calculated from the difference of sample weight before and after drying. The arithmetic mean of three parallel repetitions for each sample was taken as the final result.

3.2. Determination of Protein Content

The determination of protein content was performed in all samples of milk powder and cocoa liquor.

The protein content was determined by the Kjeldahl method. The principle of the method consisted in mineralization of samples with concentrated sulphuric acid (VI) in the presence of a catalyst (selenium-copper mixture) in the Buchi 426 Dugestion Unit. Under these conditions, the protein nitrogen was converted to an ammonium ion, which after alkalization was distilled in the form of ammonia and bound in excess boric acid in Buchi's distillation unit B-316. The ammonia solution was determined by potentiometric titration with 0.1 N hydrochloric acid standard solution. The nitrogen content in the sample was calculated on the basis of the volume of 0.1 N of hydrochloric acid standard solution used for titration, knowing that 1 cm^3 of 0.1 N of hydrochloric acid solution corresponds to 0.0014 g of nitrogen. Nitrogen was converted into protein using a multiplier calculated from the average nitrogen content in proteins, which is 16% (100:16 = 6.25). The determination was carried out in three consecutive repetitions.

3.3. Determination of Total Polyphenols Content by the Folin–Ciocalteu Method

Total polyphenol content was determined by Folin–Ciocalteu's method [31,32]. Based on preliminary tests, a 70% acetone solution was used as a solvent to prepare extracts. The extracts were prepared by weighing about 5 g of crushed test material into 300 mL grinding conical flasks and adding 100 mL of 70% acetone (*v/v*). The samples were then shaken for 30 min in a Multi-Shaker PSU 20 Biosan shaker. Following this procedure, the solutions were filtered through the corrugated filters into 100 mL grinding flasks. In order to determine the total polyphenol content, 300 μL of the extract was taken from the tubes, and 4.15 mL of deionized water, 500 μL of 20% sodium carbonate solution, and 50 μL of the Folin–Ciocalteu reagent were added. The blank sample was prepared by sampling: 300 μL of the extraction solution, 4.15 mL of deionized water, 500 μL of the 20% sodium carbonate solution, and 50 μL of the Folin–Ciocalteu reagent. Absorbance was measured at 700 nm on a Shimadzu UV-160A spectrophotometer. The apparatus was zeroed to a blank. In order to calculate the total polyphenol content, a standard curve was prepared. The standard curve was plotted for chlorogenic acid for various concentrations (0, 25, 50, 75, and 100 μL) used in absorbance measurements. Based on the results obtained, the graphical dependence of the absorbance of the solution on the amount of gallic acid contained in it was plotted. The total polyphenol content was calculated on the basis of the calibration curve and expressed in gallic acid equivalent in mg per 100 g d.m. (mg GAE·100 g^{-1} of the product Two extracts from each mass were made and polyphenols content was determined in three parallel repetitions for each extract. The average of six repetitions for each mass was considered the final result.

3.4. Determination the Ability of Extracts to Inactivate Stable DPPH Radicals

The extracts were prepared by weighing 5 g test material to conical flasks (300 mL) and adding 100 mL of 70% acetone (*v/v*). The samples were then shaken for 30 min in a Multi-Shaker PSU 20 Biosan shaker. Following that, the solutions were filtered through the corrugated filters into 100 mL flasks. Acetone extract (1 mL), acetone solution (3 mL), and added DPPH solution (1 mL) were taken to determine the appropriate sample. Acetone solution (4 mL) and DPPH solution (1 mL) were collected for the control sample. The samples were mixed and left to stand for 30 min, then absorbance was measured on the NOVASPEC II Pharmacia spectrophotometer (apparatus were zeroed for the blank test) at a wavelength of 517 nm in glass cuvettes with a diameter of 1 cm [33,34].

The antioxidant activity of the extracts against DPPH was calculated using the formula:

$$\text{Act.} = [(\text{Ak} - \text{As})/\text{Ak}] \times 100\% \tag{1}$$

where Act.—antioxidant activity (%); Ak—the absorbance of the control sample; and As—absorption of the specific sample.

3.5. The Statistical Analysis

The statistical analysis of results was performed in the Statistica 13.0. Program by using one- and two-factor analysis of variance at significance level $\alpha = 0.05$ to determine differences between the content of polyphenolic compounds and the antioxidant ability the test samples of chocolate milk masses, taking into account the effect of the protein content of milk powder. Significant differences between means were determined through Tukey's tests. The correlation matrix was analyzed to determine the relationship between polyphenol content and antioxidant activity.

4. Results and Discussion

4.1. Selection of Raw Materials for Chocolate Milk Masses Production

The selection of components differentiating the mixtures was carried out on the basis of their own characteristics and the results of analyses to which they were subjected (Tables 1 and 4).

Table 4. Cocoa liquor—content: dry matter, protein, polyphenols, and antioxidant activity (code explanation in Table 2).

Sample Code	Dry Matter	Proteins	Polyphenols	Antioxidant Activity
	% (*w/w*)	**% (*w/w*)**	**mg GAE 100 g^{-1}**	**% (*w/w*)**
Ch_1	97.75 ± 0.3	13.62 ± 1.2	3284.4 ± 20.3	93.5 ± 2.4
Ch_2	98.92 ± 0.6	14.67 ± 1.7	2881.3 ± 29.5	91.6 ± 3.6
Ch_3	98.96 ± 0.4	14.25 ± 1.6	2723.6 ± 18.6	90.2 ± 4.3

For preparation of milk masses three cocoa liquor were used: Ch_1, Ch_2, and Ch_3 (Table 4) and seven of tested powdered milk: Ml_1, Ml_2, Ml_3, Ml_4, Ml_5, Ml_6, and Ml_7 (Tables 1 and 2). The remaining part of the composition consisted of ingredients such as cocoa fat, sugar, whey, and lecithin.

The protein content in the tested cocoa masses oscillated around 13.6–14.6% (*w/w*). The values obtained in this study are also presented by the studies of Jumnongpon, et al. [35] and Torres-Moreno, et al. [36], who determined the protein content in cocoa beans from which cocoa mass is produced at the level of 13–20% (*w/w*). Selected pulp differed in producer, origin, and type of grain from which they were prepared, while in the case of dry substance and protein content they had similar parameters (Table 4).

Proteins are important bioactive food ingredients, but the potential biological activity of oligopeptides found in cocoa is insufficiently researched. Biologically active or functional proteins come primarily from food and have, in addition to their nutritional value, a physiological effect on the body [37].

In cocoa, peptides are formed naturally during the fermentation of cocoa beans and are considered important flavor precursors. The peptides come from two main protein fractions—globules, consisting of a storage protein resembling vicilin and albumin, which exhibit trypsin inhibitory properties. Cocoa proteins during natural cocoa fermentation are split into hydrophilic and hydrophobic peptides as well as amino acids by autolysis of two endogenous enzymes: aspartic endoprotease and carboxypeptidase activated by microbial metabolites such as acetic acid [38]. Milk proteins include complex groups of proteins: caseins (αs1-, αs2-, β-, and κ- casein), whey proteins (β-lactoglobulin, α-lacto-albumin, albumin serum, and immunoglobulin), and protein envelopes of fat globules. Structural differences within milk protein molecules affect their properties [39]. It is assumed that β-casein is the most effective stabilizer among milk proteins as it reduces the surface tension most [40].

According to Christian Vásquez, et al. [41] milk proteins can act as surfactants and stabilizers that are able to induce the formation of spherical micellar aggregates in the chocolate structure. Moreover, spray drying technology can induce changes in proteins that affect hydrophobicity, solubility, and denaturation [42–44]. Milk selected for preparation of milk chocolate masses contained from 11.61 to 31.1 mg of protein in 100 g of powder (Table 1). The selection was made in such a way that protein content in the analyzed milk was differentiated. This allowed for comparative analysis and evaluation of the dependence between phenol potential and protein content.

Antioxidants prevent formation of free radicals or support their removal. The milk contains several antioxidant factors such as vitamins and enzymes. Possible antioxidant activity of milk proteins and their hydrolysates has also been shown. It has been reported that peptides produced from milk protein digestion have an antioxidant effect. According to Pihlanto [45], antioxidant peptides derived from milk consist of 5–11 amino acids, including hydrophobic amino acids, proline, histidine, tyrosine, or tryptophan in the sequence.

The cocoa liquor and milk used in the study were subjected to a technological process to obtain 21 chocolate milk masses, which were tested for polyphenols content and antioxidant activity against free radicals of DPPH.

4.2. Results of Polyphenols Content Determination in Chocolate Milk Masses

The presence of polyphenols in milk depends on the animal's diet and may affect the preservative effect of milk components [46].

In milk masses made on the basis of cocoa liquor obtained from unroasted beans from Peru (Ch_1) and seven milk/milk powder mixes, the polyphenols content was obtained from about 1525 to about 1685 mg calculated as gallic acid in 100 g of product (Figure 1). The highest content of phenolic compounds was achieved by the T-5 sample with the addition of milk Ml_5 (protein content 13.19%) and the lowest by T1 based on whole milk powder Ml_1 (protein content 30.35%). The differences in the content of polyphenols were statistically significant, as indicated by five homogeneous groups.

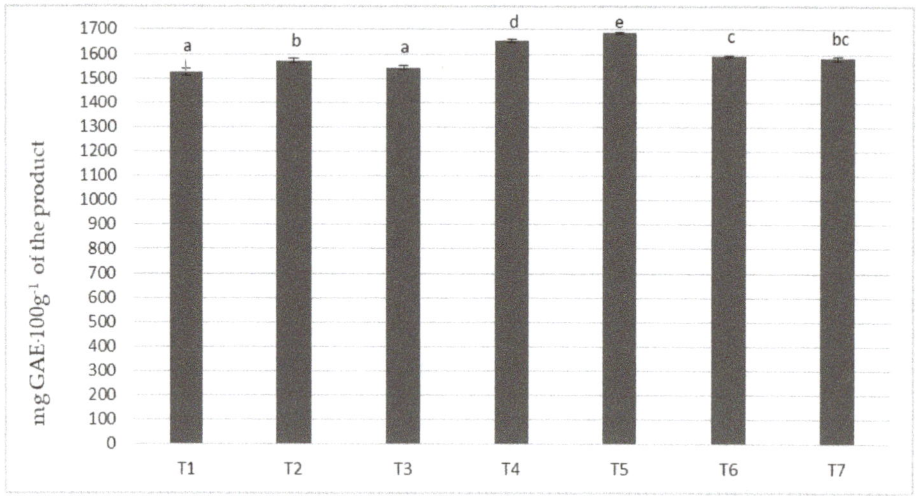

Figure 1. Polyphenols content in chocolate milk masses prepared from cocoa liquor from Peru (the same letter means that there are no statistically significant differences between the analyzed products at a confidence level of $\alpha = 0.05$; the abbreviations used in the graph are described in Table 3).

In the investigated milk masses prepared on the basis of cocoa liquor bean from the Ivory Coast (Ch_2) in combination with seven different milk powders the results of polyphenols content on the level from about 838 to almost 1071 mg per 100 g of product were obtained (Figure 2).

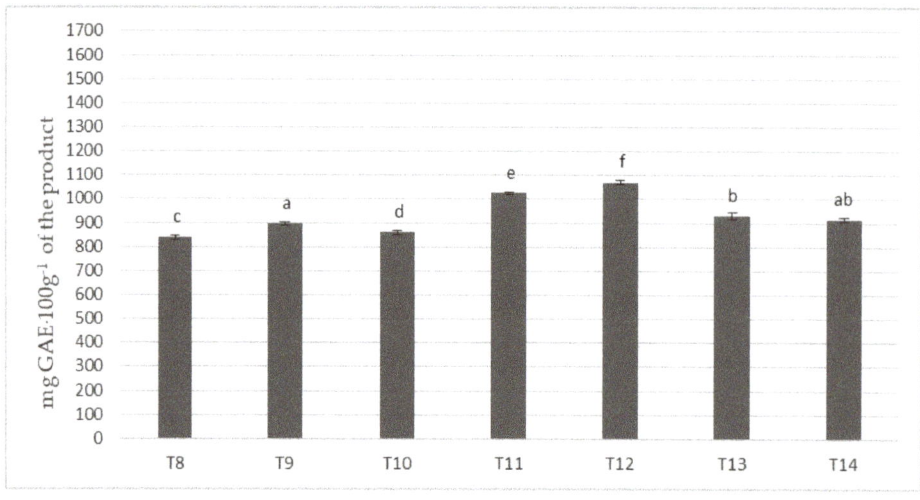

Figure 2. Polyphenols content in chocolate masses prepared from Ivory Coast cocoa liquor (the same letter means that there are no statistically significant differences between the analyzed products at a confidence level of $\alpha = 0.05$; the abbreviations used in the graph are described in Table 3).

The highest content of polyphenols was determined for the mass of T-12, to which the milk mixture Ml_5 (dried by cylinders method) was used, whereas the lowest for the mass of T-8, to which spray-dried whole milk was applied (Ml_1). All masses formed six homogeneous groups, which prove their statistically significant differentiation in terms polyphenols content. Only T-14 masses did not differ significantly from T-9 and T-13 samples. In the presented values a specific order of increase in the content of phenolic compounds from T-8, through T-10, T-9, T-14, T-13, T-11, to T-12 chocolate was observed. It is worth emphasizing that T4 and T5, as well as T11 and T12 masses were characterized by the highest content of polyphenols, and Ml_4 (protein content 18.9%) and Ml_5 (protein content 13.19%) milk with the lowest protein content was used to produce them.

Figure 3 shows the results of the determination of polyphenols content in masses obtained with the use of cocoa liquor made from cocoa beans grown in Ghana (Ch_3). Invariably, the mass with the highest content of the determined compounds remained the one with the addition of milk Ml_5 (T5 and T12 on Figures 1 and 2) and the one with the lowest one with the addition of whole milk powder Ml_1. A similar tendency to the polyphenol content was found as in the samples with the addition of cocoa liquor from the Ivory Coast. However, the differences between the masses were much less pronounced, but nevertheless formed six homogeneous groups. By comparing the results of the masses in Figure 3 with those produced from Ch_2 cocoa liquor (Ivory Coast), an average of 3–5% higher total polyphenol content could be observed in each sample in favor of those produced from beans from Ghana, regardless of the milk used. However, it is noteworthy that Ghanaian cocoa liquor beans were found to have a lower polyphenol content of approximately 6% compared to the values determined for Ivory Coast cocoa liquor bean. However, the protein content of the Ch_2 cocoa liquor was about 3% higher than that of the Ch_3 cocoa liquor (Table 4). The results obtained may confirm the literature data describing interactions between polyphenols and proteins present in the chocolate matrix. This is mainly due to the induction by proteins of the formation of spherical micellar aggregates, which close polyphenolic compounds limiting their availability.

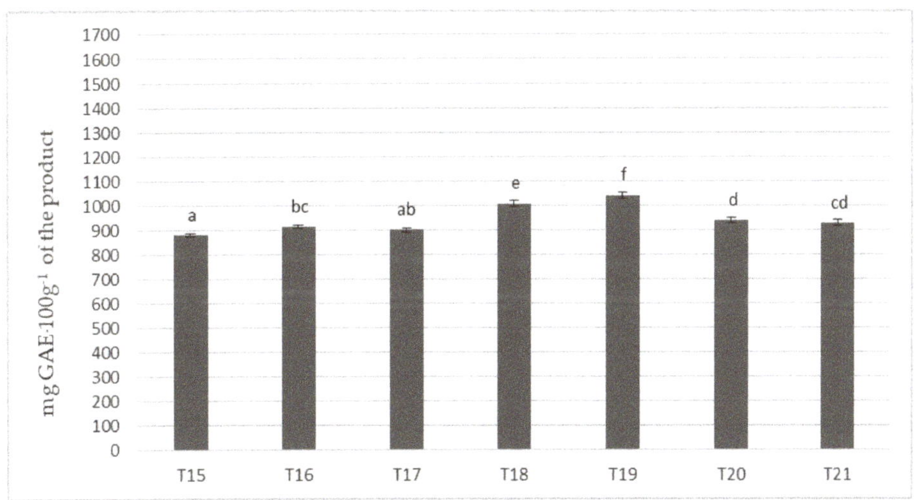

Figure 3. Polyphenols in chocolate milk masses prepared from Ghana cocoa liquor (the same letter means that there are no statistically significant differences between the analyzed products at a confidence level of $\alpha = 0.05$; the abbreviations used in the graph are described in Table 3).

Analyzing the results obtained for all 21 chocolate milk masses, it was noted that the amount of polyphenols contained in them oscillated between 35% and 50% of the values obtained for the cocoa liquor from which they were obtained. This is mainly due to the recipe in which cocoa mass constitutes less than 2/3 of the raw material composition of the finished chocolate milk mass. Additionally, not only the cocoa liquor (although certainly in the majority) brings an additional antioxidant character to the chocolate milk masses, because, as indicated by the studies of Ertan, et al. [47], whole milk powder may contain from about 50 to even 100 mg of polyphenols converted into gallic acid in 100 g of product. Summarizing these values in the finished product, which is a milk mass, it is possible to explain a polyphenol content of 1000–1500 mg in chocolates based on roasted cocoa liquor bean and correspondingly more in chocolates prepared from unroasted cocoa liquor bean. While the values for dairy products based on unroasted cocoa liquor bean were within the assumed range, this was not the case for roasted cocoa liquor bean. This could result in a small extent from the treatments (in the process of conching) to which the milk mass was subjected, losing between 5% and 10% of phenolic compounds as indicated by Di Mattia, et al. [1]. However, in the majority of cases, this loss was probably the result of possible interactions between polyphenols and proteins contained in milk and whey powder, intensified mixing time to which chocolate masses are subjected. As Gallo, et al. [48] points out, polyphenol–protein interactions may occur in milk chocolates, resulting in compounds that are inaccessible to the components of the reaction, while limiting their availability and in vitro oxidative potential. These studies showed a 15% decrease in the assay of polyphenols combined with casein, a 25% decrease in combination with whey proteins and almost 40% decrease in combination with pure ß-lactoglobulin (the main whey protein of milk). The results of these studies may explain the trends obtained in the studies carried out in this study, considering that the results obtained after statistical analysis confirmed a correlation of −0.66 between protein and polyphenol content in chocolate milk masses.

The comparative analysis of polyphenol content in all analyzed milk chocolate masses was 13 homogeneous groups, which indicates a large diversity of tested samples (Table 5). Considering that the protein content in chocolate pulp used to make the masses was at a similar level, and every third sample contained the addition of the same milk, such a large number of homogeneous groups indicate the complexity of factors shaping the antioxidant potential.

Table 5. Results of statistical analysis of polyphenols content and antiradical activity of all analyzed milk chocolate masses (code explanation in Table 3).

Codes of Masses	Polyph.	DPPH	Codes of Masses	Polyph.	DPPH	Codes of Masses	Polyph.	DPPH
T-1	J	h	T-8	A	a	T-15	B C	b
T-2	K	j	T-9	C G	c	T-16	D E	de
T-3	J	ij	T-10	A B	b	T-17	C D	c
T-4	L	m	T-11	G H	gh	T-18	G	ij
T-5	M	n	T-12	I	i	T-19	H	k
T-6	K	l	T-13	E F	de	T-20	F	fg
T-7	K	k	T-14	D E	cd	T-21	E F	ef

The same letter A, B or a, b means that there are no statistically significant differences between the analyzed products at a confidence level of α = 0.05; the abbreviations used in the graph are described in Table 3.

4.3. Results of Determination of Antioxidant Activity in Chocolate Milk Masses

Whey proteins may have an antioxidant activity, while heat treatment at very high temperature has no clear effect on the antioxidant potential of milk [49].

Antioxidant potential (AP) is an important nutritional property of food, as increased oxidative stress has an effect on most diet-related chronic diseases. In dairy products the protein fraction shows antioxidant activity, especially casein. Pihlanto study [45] reported that peptides produced from milk protein digestion have an antioxidant effect. Antioxidant peptides derived from milk consist of 5–11 amino acids, including hydrophobic amino acids, proline, histidine, tyrosine, and tryptophan in sequence.

The milk contains several antioxidants: naturally occurring vitamins (i.e., E and C), β-carotene, enzymatic systems, serum albumin, and lactoferrin, which act as chelating agents, iron-binding glycoprotein, and the activity of free radical scavenging by amino acids such as tyrosine and cysteine [45]. Therefore, polyphenols may be the main contributor to the antioxidant capacity of the food matrix [44].

The solubility of milk protein in water may be closely related to the oxidative status as oxidation improves interactions and protein aggregation [50], which reduces the solubility of milk powder [51].

In these studies, the antioxidant activity remained in the same upward trend as polyphenols. The results of the phenolic potential expressed as antioxidant activity in the masses obtained from unroasted cocoa beans are shown in Figure 4. However, it was noticeable, as in the case of the determination of phenolic compounds, that higher values, on average by 8–10%, were obtained for the masses obtained from unroasted cocoa beans. The obtained values were subjected to statistical analysis, on the basis of which the masses were assigned to seven homogeneous groups. No significant differences were observed between T2 and T3 masses, prepared successively from whole milk powder Ml_2 and Ml_3, containing 27.44% and 29.14% (*w/w*) protein, respectively.

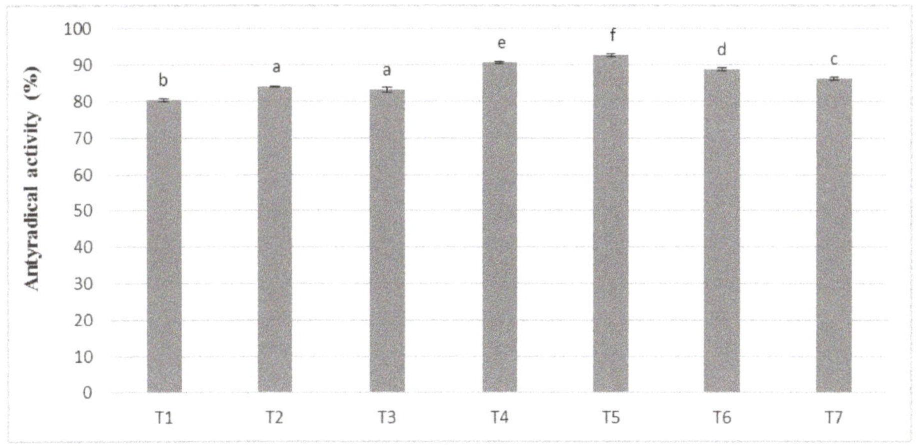

Figure 4. Antioxidant activity in chocolate milk masses prepared from unroasted cocoa liquor from Peru (the same letter means that there are no statistically significant differences between the analyzed products at a confidence level of $\alpha = 0.05$; the abbreviations used in the graph are described in Table 3).

In chocolate milk masses prepared from unroasted cocoa liquor bean from Peru, the results of determination of antioxidant activity against free radicals of DPPH were 70–80% (Figure 4). The highest activity was characterized by T5 mass with the addition of milk mixture Ml_5, while the lowest T-8 mass with addition of whole milk powder Ml_1, which coincides with the dependence obtained during polyphenols determination. Similar observations were also observed in the significance of differences in the levels of antioxidant activity as, as well as in the case of polyphenols content. Chocolate milk masses from the Ivory Coast cocoa liquor bean formed six homogeneous groups from which no significant differences between the T14 and T9 and T13 mixture were found.

In chocolate milk masses prepared from roasted cocoa liquor bean from Ivory Coast the results of determination of antioxidant activity against free radicals of DPPH were 70–80% (Figure 5). The highest activity was characterized by T12 mass with addition of milk mixture Ml_5, while the lowest T-8 mass with addition of whole milk powder Ml_1, which coincides with the dependence obtained during polyphenols determination. Similar observations were also observed in the significance of differences in the levels of antioxidant activity as well as in the case of polyphenols content. Milk masses from the Ivory Coast cocoa liquor bean formed six homogeneous groups from which no significant differences between T14 and T9 and T13 mixture resulted (Figure 5).

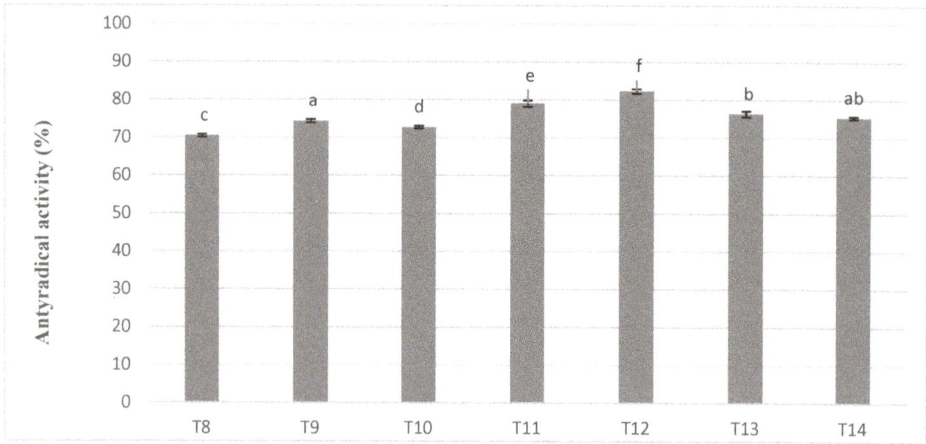

Figure 5. Antioxidant activity in chocolate milk masses prepared from Ivory Coast cocoa liquor (the same letter means that there are no statistically significant differences between the analyzed products at a confidence level of $\alpha = 0.05$; the abbreviations used in the graph are described in Table 3).

The levels of antiradical activity of milk masses prepared from cocoa liquor from Ghana also maintained very similar relationships as in the case of the previous variant and the results obtained during the determination of polyphenols content (Figure 6). Again, the mixture with the addition of milk preparation Ml_5 (protein content 13.19%) was characterized by the highest value of the properties tested (in this case antiradical activity). A similar tendency of changes in antioxidant activity as in all previous determinations were shown on Figures 5 and 6.

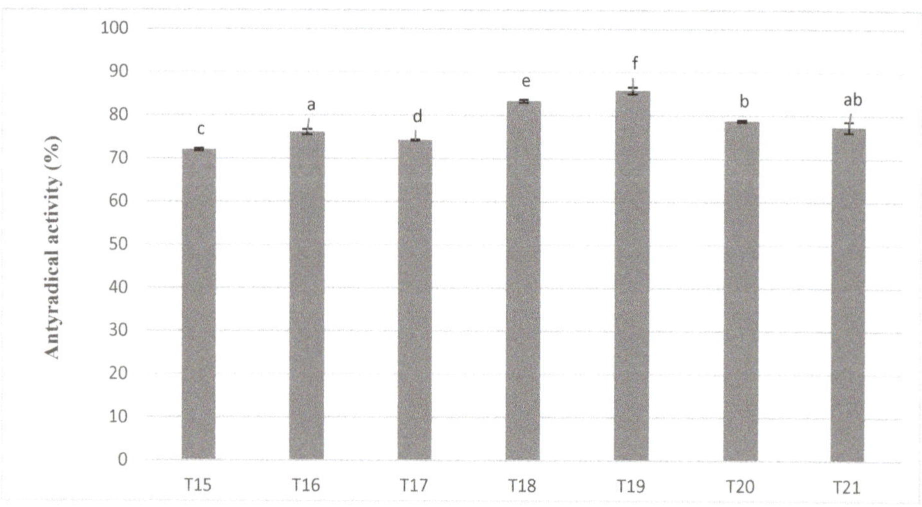

Figure 6. Antioxidant activity in chocolate milk masses prepared from cocoa liquor from Ghana (the same letter means that there are no statistically significant differences between the analyzed products at a confidence level of $\alpha = 0.05$; the abbreviations used in the graph are described in Table 3).

The antioxidant activity expressing the degree of reduction of free radicals by antioxidant compounds contained in the product is a valuable parameter of both the technological and health potential. The results obtained for all 21 milk masses allowed to determine in which of them the in vitro antioxidant potential was highest and how it changed depending on the raw materials used and the process parameters.

In the studies of Da Silva Medeiros, et al. [30], as well as Todorović, et al. [52], chocolate milk masses with declared cocoa liquor content of about 30% were distinguished by almost six times lower antioxidant activity potential, as compared to the tested bitter chocolates with declared cocoa liquor content of 67–70%. Similarly, the studies of Serafini, et al. [53] on blueberry fruit extracts and Dubeau, et al. [54] on different types of tea proved to confirm the dependence of lowering the antioxidant activity of extracts prepared after mixing them with milk. However, these values, unlike in the case of chocolates, were characterized by only 5–20% lower results compared to the activity of extracts not enriched with milk. Differences may result mainly from different protein content in milk added to tea (liquid—about 5 g/100 mL) and content of this ingredient in dried milk (about 30 g/100 g), as well as the amount of milk added in the mentioned variants.

In the investigated milk masses similar relationships were noted, because after the addition of milk or milk preparations, they reached much lower inactivation capacity of free radicals DPPH. This characteristic tendency is caused mainly by the fact, as indicated by Vertuani, et al. [55], that these masses are mixtures of various components, of which the cocoa liquor, bringing the greatest antioxidant character, is only a part of their composition. It is noteworthy, however, that this decrease was not as large as in the case of studies comparing milk chocolates to bitter chocolates. The differences were closer to those recorded for blueberry or tea extracts. This may mean that further processing and more intense industrial conditions, which were not provided for in this study, may lead to an increase in the reduction of antioxidant activity in milk chocolate. Intensified and extended by the process of tempering, the mixing of milk chocolate components causes more frequent interactions in the mixture of polyphenols with proteins, which, as indicated by Belščak, et al. [56], form complexes no longer having the same antiradical properties under the conditions of the antioxidant activity determination reaction. Similar conclusions were also drawn by Dubeau, et al. [54], who determined that casein proteins and catechin molecules in tea are mainly responsible for complex formation.

By the use of ANOVA analysis the results of antioxidant activity obtained for particular chocolate milk masses were compared to their polyphenols content. The correlation was close to 96.5%, which is a very strong correlation. It shows a simultaneous proportional decrease in antioxidant activity with a decrease in polyphenols content and its high value is also confirmed by Miller, et al. [57] in which this correlation reached almost 98%. However, more interesting from the nutritional point of view was a strong correlation between the results of protein content in masses and the content of previously discussed polyphenols or antioxidant activity. Negative correlation at the level of 0.66 and 0.76, respectively, showed a gradual proportional decrease in polyphenols and antioxidant activity with increasing protein content in the mass. The possible formation of protein–polyphenol complexes, supported by the study results and quoted in the literature, may explain this relationship. It can also be a valuable indication for producers of milk chocolate, who, while making chocolate products, also want to ensure pro-healthy values by preserving as many polyphenol compounds as possible. This relationship may also explain the observations of Serafini, et al. [53], who stated in their research that the reduction of antioxidant potential may also be influenced by the fat content, because taking into account whole, semiskimmed and skimmed milk, they obtained for blueberry fruit extracts a similarly characteristic decrease in antioxidant activity with increasing fat content. However, this dependence may have been largely due to the decrease in potential due to the antioxidant products of fat oxidation as well as to the decreasing antioxidant potential due to their combination with milk proteins.

Statistical inference was carried out with a 95% confidence level, to which the results of antioxidant activity of all tested milk chocolate masses were subjected. Fourteen homogeneous groups were obtained—one group more than for the analysis of polyphenols content (Table 5). Noteworthy is the analysis of masses T-7, T-8, T-1, T-12, T-16, and T-21, which were in the same groups in both the content of polyphenols and DPPH. A similar inference arises, as described in the analysis of polyphenol content. The analytical analysis confirmed the large variation between the results obtained, despite the fact that the tested masses can be grouped into three areas in terms of the used milk chocolate mass and in seven groups in terms of the used milk powder. The results indicate the complexity of aspects related to antioxidant potential.

Regardless of the decreasing antioxidant reduction potential in vitro studies, in many works, among others, Serafini, et al. [53], Loffredo, et al. [58], and Di Mattia, et al. [1], the same trend is not observed in studies on living organisms. In all these studies, the results of the increase in the occurrence of specific antioxidants in plasma consumed in the vicinity of milk did not differ significantly from those of milk-free counterparts. Such dependence may mainly result from the conditions of the digestive process of living organisms and many transformations and changes of substances that may occur as a result. Decreases in the amount determined in the body of epicatechin patients were recorded in studies by Neilson, et al. [59] after consumption of cocoa products containing milk. In each case, however, these values, although lower, did not show statistically significant differences. This phenomenon was explained by the variability of intestinal capture of food components and their subsequent transport between cells in patients' organisms. It follows that lowering the antioxidant character of cocoa in a product with the addition of milk, such as milk chocolate, may be of primarily technological importance.

5. Conclusions

In this study, the highest content of polyphenols was determined in chocolate milk masses obtained from milk with the lowest protein content. This relationship is least evident in the masses obtained from cocoa liquor from Ghana, which also contained the least protein. Negative correlation between protein and polyphenol content in chocolate masses was observed.

A strong correlation was also shown for the decreasing ability of chocolate milk masses to scavenge DPPH radicals with increasing protein content of milk or milk preparations used to produce them. This correlation proves that the antioxidant potential of ready-made chocolate milk masses decreases with increasing content of milk proteins and consequently their ability to neutralize and inhibit the oxidation reaction of fats contained in chocolates decreases.

A strong correlation between the content of polyphenols in chocolate milk masses prepared for testing and their antioxidant activity was shown, which proves a proportional preservation of both values in the subject of the study.

In order to confirm this hypothesis, further studies should focus on the determination of fat oxidation products in masses and in vitro tests on the analyzed analytical matrix.

Author Contributions: B.U. conceived the idea, resources, data curation, compiled the information and drafted the manuscript, edited the article for submission; T.S. resources, data curation; H.K. visualization, checking the correctness of the language, revised the article for submission; J.K. designed the study, methodology, provided guidance and revised the article for submission, supervision. All authors contributed in revision of manuscript. All authors read and approved the final manuscript.

Funding: This research received no external funding.

Conflicts of Interest: The authors declare no conflict of interest.

References

1. Di Mattia, C.D.; Sacchetti, G.; Mastrocola, D.; Serafini, M. From cocoa to chocolate: The impact of processing on in vitro antioxidant activity and the e ects of chocolate on antioxidant markers in vivo. *Front. Immunol.* **2017**, *8*, 1207. [CrossRef] [PubMed]

2. Żyżelewicz, D.; Budryn, G.; Oracz, J.; Antolak, H.; Kregiel, D.; Kaczmarska, M. The effect on bioactive components and characteristics of chocolate by functionalization with raw cocoa beans. *Food Res. Int.* **2018**, *113*, 234–244. [CrossRef] [PubMed]

3. Montagna, M.T.; Diella, G.; Triggiano, F.; Caponio, G.R.; Giglio, O.D.; Caggiano, G.; Ciaula, A.D.; Portincasa, P. Chocolate, "food of the gods": History, science, and human health. *Int. J. Environ. Res. Public Health* **2019**, *16*, 4960. [CrossRef] [PubMed]

4. Zimmermann, B.F.; Ellinger, S. Cocoa, chocolate, and human health. *Nutrients* **2020**, *12*, 698. [CrossRef] [PubMed]

5. Mota-Gutierrez, J.; Barbosa-Pereira, L.; Ferrocino, I.; Cocolin, L. Traceability of functional volatile compounds generated on inoculated cocoa fermentation and its potential health benefits. *Nutrients* **2019**, *11*, 884. [CrossRef] [PubMed]

6. Tsang, C.; Hodgson, L.; Bussu, A.; Farhat, G.; Al-Dujaili, E. Effect of polyphenol-rich dark chocolate on salivary cortisol and mood in adults. *Antioxidants* **2019**, *8*, 149. [CrossRef] [PubMed]

7. Oracz, J.; Zyzelewicz, D. In vitro antioxidant activity and ftir characterization of high-molecular weight melanoidin fractions from different types of cocoa beans. *Antioxidants* **2019**, *8*, 560. [CrossRef]

8. Da Veiga Moreira, I.M.; De Figueiredo Vilela, L.; Santos, C.; Lima, N.; Schwan, R.F. Volatile compounds and protein profiles analyses of fermented cocoa beans and chocolates from different hybrids cultivated in Brazil. *Food Res. Int.* **2018**, *109*, 196–203. [CrossRef]

9. Engeseth, N.J.; Ac Pangan, M.F. Current context on chocolate flavor development—A review. *Curr. Opin. Food Sci.* **2018**, *21*, 84–91. [CrossRef]

10. Hinneh, M.; Semanhyia, E.; van de Walle, D.; de Winne, A.; Tzompa Sosa, D.A.; Scalone, G.L.; de Meulenaer, B. Assessing the influence of pod storage on sugar and free amino acid profiles and the implications on some maillard reaction related flavor volatiles in forastero cocoa beans. *Food Res. Int.* **2018**, *111*, 607–620. [CrossRef]

11. Ramli, N.; Hassan, O.; Said, M.; Samsudin, W.; Idris, N.A. Influence of roasting conditions on volatile flavor ofroasted Malaysian cocoa beans. *J. Food Process. Preserv.* **2006**, *30*, 280–298. [CrossRef]

12. Farah, D.M.H.; Zaibunnisa, A.H.; Misnawi, J.; Zainal, S. Effect of roasting process on the concentration of acrylamide and pyrazines in roasted cocoa beans from different origins. *APCBEE Procedia* **2012**, *4*, 204–208. [CrossRef]

13. Ioannone, F.; di Mattia, C.D.; de Gregorio, M.; Sergi, M.; Serafini, M.; Sacchetti, G. Flavanols, proanthocyanidins and antioxidant activity changes during cocoa (Theobroma cacao L.) roasting as affected by temperature and time of processing. *Food Chem.* **2015**, *174*, 256–262. [CrossRef] [PubMed]

14. Stanley, T.H.; Van Buiten, B.; Baker, S.A.; Elias, R.J.; Anantheswaran, R.C.; Lambert, J.D. Impact of roasting on the flavan-3-ol com-position, sensory-related chemistry, and in vitro pancreatic lipase inhibitory activity of cocoa beans. *Food Chem.* **2018**, *255*, 414–420. [CrossRef]

15. Żyżelewicz, D.; Krysiak, W.; Oracz, J.; Sosnowska, D.; Budryn, G.; Nebesny, E. The influence of the roasting process conditions on the polyphenol content in cocoa beans, nibs and chocolates. *Food Res. Int.* **2016**, *89*, 918–929. [CrossRef]

16. Beckett, S.T. Liquid chocolate making. In *The Science of Chocolate*, 2nd ed.; Paperbacks; Royal Society of Chemistry: London, UK, 2008; Volume 4, pp. 61–79.

17. Owusu, M.; Petersen, M.A.; Heimdal, H. Effect of fermantation method roasting and conching conditions on the aroma volatiles of dark chocolate. *J. Food Process. Preserv.* **2012**, *36*, 446–456. [CrossRef]

18. Vivar-Vera, G.; Torrestiana Sanchez, B.; Monrey Rivera, J.A.; Brito de La Fuente, E. Influence of temperature and mixing speed on dynamic rheological properties of dark chocolate mass during conching. In *Food Science and Food Biotechnology Essentials: A Contemporary Perspective*; Moorillan, G.V.N., Ortega-Rivas, E., Eds.; Associacion Mexicana de Ciencia de los Alimentos, A.C: Durango, Mexico, 2014; pp. 247–252.

19. Toker, O.S.; Palabiyik, I.; Konar, N. Chocolate quality and conching. *Trends Food Sci. Technol.* **2019**, *91*, 446–456. [CrossRef]

20. Chen, K.; Zhao, J.; Shi, X.; Abdul, Q.; Jiang, Z. Characterization and antioxidant activity of products derived from xylose–bovine casein hydrolysate maillard reaction: Impact of reaction time. *Foods* **2019**, *8*, 242. [CrossRef]

21. Schumacher, A.B.; Brandelli, A.; Schumacher, E.W.; Macedo, F.C.; Pietai, L.; Klug, T.; de Jong, E.V. Development and evaluation of a laboratory scale conch for chocolate production. *Int. J. Food Sci. Technol.* **2009**, *44*, 616–622. [CrossRef]

22. Albak, F.; Tekin, A.R. Variation of total aroma and polyphenol content of dark chocolate during three phase of conching. *J. Food Sci. Technol.* **2016**, *53*, 848–855. [CrossRef]

23. Di Mattia, C.; Martuscelli, M.; Sacchetti, G.; Bahaydt, B.; Matsrocola, D.; Pittia, P. Effect of different conching processes on procyanidins content and antioxidant properties of chocolate. *Food Res. Int.* **2014**, *63*, 367–372. [CrossRef]

24. E.U. Directive 2000/36/EC of the European parliament and of the council of 23 June 2000 relating to cocoa and chocolate products intended. *Off. J.* **2020**, *197*, 19–25.

25. Furlán, L.T.R.; Baracco, Y.; Lecot, J.; Zaritzky, N.; Campderrós, M.E. Influence of hydrogenated oil as cocoa butter replacers in the development of sugar-free compound chocolates: Use of inulin as stabilizing agent. *Food Chem.* **2017**, *217*, 637–647. [CrossRef] [PubMed]

26. Fitzpatrick, J.; Barry, K.; Delaney, C.; Keogh, K. Assessment of the flowability of spraydried milk powders for chocolate manufacture. *Lait* **2005**, *85*, 269–277. [CrossRef]

27. Haylock, S.J.; Dodds, T.M. Ingredients from Milk. In *Industrial Chocolate Manufacture and Use*, 3rd ed.; Beckett, S.T., Ed.; Blackwell Science: Oxford, UK, 1999; pp. 1–77.

28. Kowalska, J.; Małoszewska, E. Commodity evaluation of high cocoa chocolates. *Sci. Nat. Technol.* **2009**, *3*, 141.

29. Maidannyk, V.; McSweeney, D.J.; Hogan, S.A.; Miao, S.; Montgomery, S.; Auty, M.A.E.; McCarthy, N.A. Water sorption and hydration in spray-dried milk protein powders: Selected physicochemical properties. *Food Chem.* **2020**, *304*, 125418. [CrossRef]

30. Da Silva Medeiros, N.; Marder, R.K.; Farias Wohlenberg, M.; Funchal, C.; Dani, C. Total phenolic content and antioxidant activity of different types of chocolate, milk, semisweet, dark, and soy, in cerebral cortex, hippocampus, and cerebellum of wistar rats. *Biochem. Res. Int.* **2015**, *2015*, 294659. [CrossRef]

31. Wołosiak, R.; Drużyńska, B.; Piecyk, M.; Worobiej, E.; Majewska, E.; Lewicki, P. Influence of industrial sterilisation, freezing and steam cooking on antioxidant properties of green peas and string beans. *Int. J. Food Sci. Technol.* **2011**, *46*, 93–100. [CrossRef]

32. Singleton, V.L.; Rossi, J.A. Colorimetry of total phenolics with phosphomo-lybdic-phosphotungstic acid reagents. *Am. J. Enol. Viticult.* **1965**, *16*, 144–158.

33. Brand-Williams, W.; Cuvelier, M.E.; Berset, C. Use of a free radical method to evaluate antioxidant activity. *LWT* **1995**, *28*, 25–30. [CrossRef]

34. Tailor, C.S.; Goyal, A. Activity by DPPH radical scavenging method of ageratum conyzoideslinn. *Leaves Am. J. Ethnomed.* **2014**, *1*, 244–249.

35. Jumnongpon, R.; Chaiseri, S.; Hongsprabhas, P.; Healy, J.P.; Meade, S.J.; Gerrard, J.A. Cocoa protein crosslinking using Maillard chemistry. *Food Chem.* **2012**, *134*, 375–380. [CrossRef]

36. Torres-Moreno, M.; Tarrega, A.; Costell, E.; Blanch, C. Dark chocolate acceptability: Influence of cocoa origin and processing conditions. *J. Sci. Food Agric.* **2012**, *92*, 404–411. [CrossRef] [PubMed]

37. Hebert, E.M.; Saavedra, L.; Ferranti, P. Bioactive peptides derived from casein and whey proteins. In *Biotechnology of Lactic Acid Bacteria: Novel Application*; Mozzi, F., Raya, R.R., Vignolo, G.M., Eds.; Blackwell Publishing: Oxford, UK, 2010. [CrossRef]

38. Marseglia, A.; Dellafiora, L.; Prandi, B.; Lolli, V.; Sforza, S.; Cozzini, P.; Tedeschi, T.; Galaverna, G.; Caligiani, A. Simulated gastrointestinal digestion of cocoa: Detection of resistant peptides and in silico/in vitro prediction of their ace inhibitory activity. *Nutrients* **2019**, *11*, 985. [CrossRef]

39. Darewicz, M.; Dziuba, J. Structure and functional properties of milk proteins. *Food Sci. Technol. Qual.* **2005**, *2*, 47–60.

40. Domian, E. Profile of spray-dried emulsions stabilised by milk proteins. *Food Sci. Technol. Qual.* **2011**, *18*, 5–23. [CrossRef]

41. Vásquez, C.; Henríquez, G.; López, J.; Penott-Chang, E.; Sandoval, A.; Müller, A. The effect of composition on the rheological behavior of commercial chocolates. *LWT* **2019**, *111*, 744–750. [CrossRef]

42. Singh, H.; Creamer, L.K. Denaturation, aggregation and heat stability of milk protein during the manufacture of skim milk powder. *J. Dairy Res.* **1991**, *58*, 269–283. [CrossRef]

43. Kurozawa, L.E.; Park, K.J.; Hubinger, M.D. Effect of carrier agents on the physicochemical properties of a spray dried chicken meat protein hydrolysate. *J. Food Eng.* **2009**, *94*, 326–333. [CrossRef]

44. Scheidegger, D.; Radici, P.M.; Vergara-Roig, V.A.; Bosio, N.S.; Pesce, S.F.; Pecora, R.P.; Romano, J.C.P.; Kivatinitz, S.C. Evaluation of milk powder quality by protein oxidative modifications. *J. Dairy Sci.* **2013**, *96*, 3414–3423. [CrossRef]

45. Pihlanto, A. Antioxidative peptides derived from milk proteins. *Int. Dairy J.* **2016**, *16*, 1306. [CrossRef]

46. Besle, J.M.; Viala, D.; Martin, B.; Pradel, P.; Meunier, B.; Berdagué, J.L.; Fraisse, D.; Lamaison, J.L.; Coulon, J.B. Ultraviolet-absorbing compounds in milk are related to forage polyphenols. *J. Dairy Sci.* **2010**, *93*, 2846–2856. [CrossRef] [PubMed]

47. Ertan, K.; Bayana, D.; Gokce, O.; Alatossava, J.T.; Yilmaz, Y.; Gursoy, O. Total antioxidant capacity and phenolic content of pasteurized and UHT-treated cow milk samples marketed in Turkey. *Akad. Gıda Derg.* **2017**, *15*, 103–108. [CrossRef]
48. Gallo, M.; Vinci, G.; Graziani, G.; de Simone, C.; Ferranti, P. The interaction of cocoa polyphenols with milk proteins studied by protomic techniques. *Food Res. Int.* **2013**, *54*, 406–415. [CrossRef]
49. Arranz, E.; Corrochano, A.R.; Shanahan, C.; Villalva, M.; Jaime, L.; Santoyo, S.; Callanan, M.J.; Murphy, E.; Giblin, L. Antioxidant activity and characterization of whey protein-based beverages: Effect of shelf life and gastrointestinal transit on bioactivity. *Innov. Food Sci. Emerg.* **2019**, *57*. [CrossRef]
50. Scheidegger, D.; Pecora, R.P.; Radici, P.M.; Kivatinitz, S.C. Protein oxidative changes in whole and skim milk after ultraviolet or fluorescent light exposure. *J. Dairy Sci.* **2010**, *93*, 5101–5109. [CrossRef]
51. Thomas, M.E.C.; Scher, J.; Desobry-Banon, S.; Desobry, S. Milk powders ageing: Effect on physical and functional properties. *Crit. Rev. Food Sci. Nutr.* **2004**, *44*, 297–322. [CrossRef]
52. Todorovic, V.; Radojcic, I.; Todorovic, Z.; Jankovic, G.; Dodevsk, M.; Sobajic, S. Polyphenols methylxanthines and antioxidant capacity of chocolates produced in Serbia. *J. Food Compos.* **2015**, *41*, 137–143. [CrossRef]
53. Serafini, M.; Testa, M.F.; Villaño, D.; Pecorari, M.; Peluso, I.; Brambilla, A. Antioxidant activity of blueberry fruit is impaired by the association with milk. *Ann. Nutr. Metab.* **2008**, 55. [CrossRef]
54. Dubeau, S.; Samson, G. Dual effect of milk on the antioxidant capacity of green, Darjeeling, and English breakfast teas. *Food Chem.* **2010**, *122*, 539–545. [CrossRef]
55. Vertuani, S.; Scalambra, E.; Vittorio, T.; Bino, A.; Malisardi, G.; Baldisserotto, A.; Manfredini, S. Evaluation of antiradical activity of different cocoa and chocolate products: Relation with lipid and protein composition, 2014. *J. Med. Food* **2013**, *17*, 512–516. [CrossRef] [PubMed]
56. Belščak, A.; Komes, D.; Horzic, D.; Kovacevic, K.; Karlovic, D. Comparative study of commercially available cocoa products in terms of their bioactive composition. *Food Res. Int.* **2009**, *42*, 707–716. [CrossRef]
57. Miller, K.B.; Stuart, D.A.; Smith, N.L.; Lee, C.Y.; Mchale, N.L.; Flanagan, J.A.; Ou, B.; Hurst, W.J. Antioxidant activity and polyphenol and procyanidin contents of selected commercially available cocoa-containing and chocolate products in the United States. *J. Agric. Food Chem.* **2006**, *54*, 4062–4068. [CrossRef] [PubMed]
58. Loffredo, L.; Perri, L.; Nocella, C.; Violi, F. Antioxidant and antiplatelet activity by polyphenol-rich nutrients: Focus on extra virgin olive oil and cocoa. *Br. J. Clin. Pharmacol.* **2017**, *83*, 96–102. [CrossRef] [PubMed]
59. Nielsen, I.; Bentsen, I.; Lisby, M. A Flp-nick system to study repair of a single protein-bound nick in vivo. *Nat. Methods* **2009**, *6*, 753–757. [CrossRef]

Article

Food-Safe Process for High Recovery of Flavonoids from Cocoa Beans: Antioxidant and HPLC-DAD-ESI-MS/MS Analysis

Said Toro-Uribe [1], Elena Ibañez [2], Eric A. Decker [3], Arley René Villamizar-Jaimes [4] and Luis Javier López-Giraldo [1,*]

[1] School of Chemical Engineering, Food Science and Technology Research Center (CICTA), Universidad Industrial de Santander, Carrera 27, Calle 9, Bucaramanga 68002, Colombia; saidtorouribe@gmail.com
[2] Foodomics Laboratory, Institute of Food Science Research (CIAL, CSIC-UAM), Nicolás Cabrera 9, 28049 Madrid, Spain; elena.ibanez@csic.es
[3] Department of Food Science, University of Massachusetts, Chenoweth Laboratory, 100 Holdsworth Way, Amherst, MA 01003, USA; edecker@foodsci.umass.edu
[4] School of Chemistry, Food Science & Technology Research Center (CICTA), Universidad Industrial de Santander, Carrera 27, Calle 9, Bucaramanga 68002, Colombia; arleyvil@uis.edu.co
* Correspondence: ljlopez@uis.edu.co; Tel.: +57-(300)377-8801

Received: 29 February 2020; Accepted: 25 March 2020; Published: 27 April 2020

Abstract: Considering the increasing interest in the incorporation of natural antioxidants in enriched foods, this work aimed to establish a food-grade and suitable procedure for the recovery of polyphenols from cocoa beans avoiding the degreasing process. The results showed that ultrasound for 30 min with particle sample size < 0.18 mm changed the microstructure of the cell, thus increasing the diffusion pathway of polyphenols and avoiding the degreasing process. The effect of temperature, pH, and concentration of ethanol and solute on the extraction of polyphenols was evaluated. Through a 2^4 full factorial design, a maximum recovery of 122.34 ± 2.35 mg GAE/g, 88.87 ± 0.78 mg ECE/g, and 62.57 ± 3.37 mg ECE/g cocoa beans, for total concentration of polyphenols (TP), flavonoids (TF), and flavan-3-ols (TF3), respectively, was obtained. Based on mathematical models, the kinetics of the solid–liquid extraction process indicates a maximum equilibrium time of 45 min. Analysis by HPLC-DAD-ESI-MS/MS showed that our process allowed a high amount of methylxanthines (10.43 mg/g), catechins (7.92 mg/g), and procyanidins (34.0 mg/g) with a degree of polymerization >7, as well as high antioxidant activity determined by Oxygen Radical Absorbance Capacity (1149.85 ± 25.10 μMTrolox eq/g) and radical scavenging activity (DPPH•, 120.60 ± 0.50 μM Trolox eq/g). Overall, the recovery method made possible increases of 59.7% and 12.8% in cocoa polyphenols content and extraction yield, respectively. This study showed an effective, suitable and cost-effective process for the extraction of bioactive compounds from cocoa beans without degreasing.

Keywords: cocoa; polyphenols; solid–liquid kinetic extraction; antioxidants

---

1. Introduction

Flavanols are the most abundant substances of polyphenols found in cocoa, with a degree of polymerization ranging from monomers to polymeric proanthocyanidins [1]. Therefore, unfermented cocoa bean is composed of 1.3–3.3% methylxanthines [2] and of about 6% condensed flavan-3-ols [3]; as a result, it is has been listed as the 4th richest dietary source of polyphenols [4]. Cocoa is one of the top crops in Colombia in terms of economic impact, with a national production record of 59,665 tons in 2019 according to National Cacao Producers Federation. There is also a government initiative aiming at a switch from cocaine to cocoa and to join forces with the private sector to enhance Colombia's competitiveness at an international level, which has been established in the comprehensive national program for the replacement of illegal crops [5].

Polyphenols have gained increasing attention as supplements and additives in functional foods due to their nutraceutical properties and beneficial health properties. The recovery of polyphenols from cocoa and its by-products through several technologies such as maceration [6], microwave [7], and pressurized liquid extraction [8] has been previously reported. Most of them use powder cocoa beans obtained from processing operations consisting mainly of (a) drying and particle size reduction pretreatment, (b) degreasing, and (c) recovery of secondary metabolites. During the pretreatment stage, low drying temperatures and short times are preferred to avoid oxidation of flavonoids while small particle size is suggested to increase mass transfer.

The total fat content of whole cocoa beans is over 50% (on a dry basis) [9], which constitutes a barrier for the release of polyphenols from the cells; therefore, a degreasing process is commonly employed that can include pressing of the beans, or using a solvent extraction (i.e., hexane, chloroform, petroleum ether or other non-polar solvents) [10,11] to achieve a final fat content lower than < 12 wt.% [12]. When defatting using solvents, an additional step for residual solvent removal is thus needed. Supercritical fluid CO_2 can be used, which leaves no residue in the final product, but it is more expensive [13].

During the last stage, the extraction yield can vary not only due to the pre-treatment step, but also due to the type of solvent, contact time, temperature, solid to solvent ratio, the structure of the solid matrix and pH [14]. Although several solvents can be used to extract cocoa polyphenols, polar solvents approved by food regulations agencies (i.e., FDA and EFSA) for human consumption (that is, water, ethanol or a mixture of both) are preferred.

Some patented processes can be found in the literature for producing cocoa polyphenol concentrate; these include a series of sequential steps consisting on inhibition of enzymatic browning (blanching with hot water or steam), pressing and supercritical CO_2 extraction for fat removal or degreasing with hexane over 6 h [13], reduction of particle size (<500 um) using cryogenic milling (under −5 °C), and extraction of polyphenols using hot water [13], 60% 2-propanol [15], acetone/water/acetic acid [16] and aqueous methanol [17].

These are good examples showing that no single universal extraction processes can be employed for the extraction of polyphenols from different plant sources [18]. Therefore, the goal of the present work was to establish the optimal experimental conditions to enhance higher recovery of polyphenols from cocoa beans without degreasing. Thus, a suitable method that was reproducible for large-scale production (0.05–10 L), cheap, used food grade solvents and reliable for food applications was developed based on ultrasound-assisted solid-liquid extraction using aqueous ethanol. The best operational conditions for the pre-treatment process (drying and particle size reduction to avoid the degreasing) and for the recovery of polyphenols were established. In addition, chemical characterization and antioxidant activity of the cocoa polyphenol extract were also studied.

2. Material and Methods

2.1. Reagents

All the chemicals used were analytical or reagent grade and were not purified further. Folin-Ciocalteu reagent, gallic acid, sodium carbonate, ascorbic acid, L-cysteine, sodium phosphate monobasic monohydrate ($NaH_2PO_4·H_2O$), disodium hydrogen phosphate (Na_2HPO_4), caffeine, and theobromine were obtained from Sigma Aldrich (St. Louis, MO, USA). (+)-Catechin hydrate (≥99%; ASB-000003310), (−)-epicatechin (≥99%; ASB-00005127), procyanidin B2 (≥90%; ASB-00016231) were purchased from ChromaDex Inc. (Irvine, CA, USA). Acetonitrile, methanol, (HPLC-grade), ethanol, *n*-hexane, citric acid and formic acid were acquired from Merck (Merck, Germany). Deionized water (18 MΩ/cm) from an Aqua Solution system (Aqua solution, Inc. Jasper, Georgia, USA) was used for the preparation of all solutions.

2.2. Pre-Treatment of Cocoa Beans

Fresh cocoa pods (Trinitary, clone ICS 39) were collected at Villa Santa Monica (San Vicente de Chucurí, Santander, Colombia) and immediately protected from light and transported on ice to CICTA Lab for processing. Beans were removed manually from the pods using a knife, cleaned and separated from the pulp (mucilage) using a mucilage remover (Penagos Ltd.a, Colombia). After that, beans were heat-treated (96 °C for 6.4 min) with an enzyme inhibition solution (70 mM·L-ascorbic acid and L-cysteine, ratio 1:1 v/v) until reaching a polyphenol oxidase inhibition up to 93%. The inactivated beans were frozen at −20 °C and used for further analysis.

2.2.1. Drying and Milling Process

Drying and milling processes were evaluated as follows:

(a) Drying: Beans after polyphenol oxidase (PPO) inactivation were used immediately to evaluate the effect of drying technology on the total polyphenol content. To do so, beans were chopped (cross section of 50×30 mm^2). Then, beans were a) oven dried (FD 23, Binder, Germany) at both 50 °C and 70 °C at atmospheric pressure, and b) freeze-dried (Labconco Corp., Kansas City, MO, USA) at −84 °C processing temperature and 13 Pa constant pressure in the drying chamber to obtain final humidity < 4%. The moisture content was evaluated by AOAC method 931.04 (AOAC, 1990).

(b) Milling: As a strategy to avoid the use of a non-GRAS (Generally Recognized as Safe) solvent for removing the fat from cocoa beans, different particle size distributions and ultrasound time were evaluated as a function of total phenol content. Dried beans were milled at max speed at −20 °C in N$_2$ environment for 30 s during three cycles (Grindomix GM 200, Retsch, Haan, Germany). The milled samples were sieved through steel mesh (W.S. Tyler, Mentor, OH, USA) with a sieve shaker (Gilson, Screen Co., USA) and fractionated in three groups: sieved and retained on 20 to 40-mesh (sample 1); 40 to 80-mesh (sample 2); and 80 to 200-mesh (sample 3). After that, the powdered sample was immersed in 50% aqueous ethanol and ultrasonicated (35 kHz, ice bath at 4 °C, Elma, Ultrasonic LC20H, Germany) for several intervals of time. A defatted cocoa sample (<5 wt.%) was employed as a control sample. To do so, the cocoa bean powder (1.0 g) was three times defatted with n-hexane (10 mL, extraction in an ultrasonic bath at 25 °C for 15 min). The resulting powder was dried overnight at room temperature.

In all experiments, a standard polyphenol extraction step was carried out as follows: 1 g of sample was added to 60 mL of a mixture of 50% ethanol and water (w/w) at 50 °C with stirring at 300 rpm for 30 min using a magnetic stirrer hotplate (IKA C-MAG HS7, Germany) with temperature being monitored with a thermocouple (IKA ETS-D5, Germany). The resulting extract was centrifuged (5000× g, 4 °C, 20 min), then the supernatant was filtered through 0.45 μm hydrophilic PTFE filter (Millipore, Milford, USA), and the total polyphenol content was measured immediately as detailed in Section 2.4.1.

2.2.2. Scanning Electron Microscopy

Scanning electron microscopy (SEM) was used to evaluate the microstructure of (a) raw cocoa beans, (b) cocoa beans with reduced PPO activity, and (c) cocoa beans with reduced PPO and treated by ultrasound. Bean samples cut into longitudinal and transversal sections were mounted on aluminum stubs with double-sided carbon adhesive tape and observed using the XL-30 Environmental SEM (Philips, USA) at 25 kV accelerating voltage with the BSE (backscattered electron) detector. The images were stored in TIFF format at 712×484 pixel in grayscale with brightness values between 0 and 255 for each pixel constituting the image.

2.3. Solid–Liquid Extraction of Polyphenols

Extraction temperature, pH, solute/solvent ratio, and ethanol/water ratio were evaluated as the major factors that can affect the extraction yield. pH was adjusted with 1 M citric acid. The combination of these factors was modeled through a surface design consisting of 2^4 + four replicates at the central point + triplicates at the start point. Low and high levels for the different factors were, as follows: Temperature (40, 60 °C); pH (3, 5); solute/solvent ratio (*w/v*) (1/60, 1/30); ethanol/water ratio (*v/v*) (25/75, 75/25).

The resulting extracts were centrifuged at 5000× *g*, 4 °C for 15 min and the supernatant was collected and filtered using Whatman N°1 filter paper (Whatman, Inc., USA), and used for further experiments. At maximum extraction conditions, large-scale recovery (up to 10 L) controlled by Bioflow-110 bioreactor (New Brunswick Scientific, Enfield, CT, USA) was used to prove that the scale up of the process is possible.

2.4. Determination of Total Polyphenol, Total Flavonoids, and Total Flavan-3-ols Content

2.4.1. Total Polyphenol Content by Folin-Ciocalteu

The total polyphenol (TP) content was assayed using the Folin-Ciocalteu reagent according to Singleton et al. [19] with modifications as follows. The reaction was initiated by the addition of 50 μL of the sample with 1.5 mL of 10-fold diluted Folin-Ciocalteu reagent. After 5 min, 1.5 mL of 7.5% (*w/v*) sodium carbonate was added and vortexed for 10 s. The reaction medium was stored in the dark for 1 h at 25 °C. Absorbance was measured at 765 nm (Genesys 20; Thermo Scientific-Fisher, Waltham, MA, USA) against a blank sample. A calibration curve of gallic acid (ranging from 0.01–0.8 mg/mL, $r^2 = 0.99$) was prepared, and the results were expressed as mg gallic acid equivalents (GAE) per gram of dried cocoa beans.

2.4.2. Total Flavonoid Assay

Total flavonoid (TF) content was measured according to Zhishen et al. [20]. Sample (500 μL) was added to 5 mL volumetric flask containing 2 mL H_2O followed by addition of 0.15 mL 5% $NaNO_2$. After 5 min, 0.15 mL 10% $AlCl_3$ was added and 1 min after, 1 mL 1M $NaOH$ was added. The total volume was made up to 5 mL with H_2O. The reactants were mixed well and stored in the dark for 15 min at 25 °C. Absorbance was measured at 510 nm (Genesys 20; Thermo Scientific-Fisher, Waltham, MA, USA) against a blank sample. Total flavonoid content was expressed as mg (−)-epicatechin equivalents (ECE) per gram of dried cocoa beans.

2.4.3. Total Flavan-3-Ol Assay

Total flavan-3-ol (TF3) content was determined by the vanillin-H_2SO_4 assay as described by Sun [21]. The reaction consisted of 1 mL sample in methanol with 2.5 mL of 1% vanillin in methanol and 2.5 mL of 9N H_2SO_4. The reaction medium was well mixed at 30 °C and allowed to react for 15 min. Absorbance was measured at 500 nm (Genesys 20; Thermo Scientific-Fisher, Waltham, MA, USA) against a blank sample. Total flavan-3-ol content was expressed as mg (−)-epicatechin equivalents (ECE) per gram of dried cocoa beans.

2.5. Kinetic of Solid–Liquid (S-L) Extraction of Polyphenols

Once the optimum S-L extraction conditions were selected, the extraction kinetics were evaluated by plotting the concentration of the isolated target analyte versus time. Aliquots (100 μL) of each sample were taken out at various times to measure the total concentration of polyphenol, flavonoid, and flavan-3-ol. The extraction curves were adjusted by several kinetic models previously documented [22–26].

2.6. Characterization by HPLC-DAD-ESI-MS/MS

HPLC analysis was performed on a Shimadzu (LC-2030 LT Series-i, USA) equipped with a photodiode detector, solvent degasser, quaternary pump, autosampler with temperature control, and thermostat column compartment. The separation was achieved using a C18-phenyl column (4.6 × 50 mm, 2.5 μm particle size) (Xbridge®, Waters, USA) protected with a security guard obtained from Phenomenex (AJ0-8788, Phenomenex, Torrance, CA, USA). The procedure consisted of acidified water (water/formic acid, 99.9:0.01 *v/v*) (solvent A) and acidified acetonitrile (acetonitrile/formic acid, 99.9:0.01 *v/v*) (solvent B). The optimized linear gradient was as follows: 0–8 min, 2% B; 8–37 min, 10% B; 37–40 min, 0% B and 2% B for 10 min. The flow rate was 0.8 mL/min, and the temperature was 60 °C. The detector acquisition was 190–800 nm. The calibration curves for caffeine, theobromine, catechins, and dimer B2 were made from commercially available analytical standards ($r^2 = 0.99$). Oligomeric procyanidin calibration curve was performed from isolated fractions ($r^2 = 0.98$) according to Toro-Uribe et al. [27]. All the results are expressed as mg of sample per g of cocoa beans (dry matter basis).

The mass detector was an Agilent 6320 Ion Trap LC/MS (Agilent Technology, Waldbronn, Germany) equipped with an ESI source and ion trap mass analyzer which was controlled by the 6300 series trap control software (Bruker Daltonik GmbH). MS/MS analyses were carried out to obtain the structural information of the separated compounds. Mass spectrometer was operated under positive and negative ESI mode, nebulizer pressure, 40 psi; dry gas, 12 L min^{-1}; dry temperature, 350 °C, and mass spectra recorded from 90 to 2200 *m/z*. MS characterization features were analyzed using extraction ion compound tool and commercial standards. It was also consulted METLIN and HMBD databases for matching exact mass.

2.7. Antioxidant Assays

Oxygen radical absorbance capacity (ORAC) assay described by Huang et al. [28] was carried out as follow: 96 well microplates were filled with 50 μL of the daily working fluorescein solution (4×10^{-3} mM in 75 mM phosphate buffer, pH 7.4), 50 μL of samples at a known concentration, and incubated at 37 °C for 10 min in a microplate reader (Synergy HT Multi-Detection, Biotek Instruments, Inc. Winooski, VT, USA). The reaction was initiated by the addition of 50 μL of 2,2′-Azobis(2-methylpropionamidine) dihydrochloride (221 mM in 75 mM phosphate buffer, pH 7.4) and the fluorescence decay was monitored kinetically for 2 h, using emission and excitation wavelength of 485 nm and 528 nm, respectively. A calibration curve was prepared with 12.5–375 μM Trolox ($r^2 = 0.99$). Results are expressed as μM Trolox equivalents per g of cocoa beans (dry matter basis) as follows:

$$ORAC = \frac{AUC_{sample} - AUC_{blank}}{AUC_{Trolox} - AUC_{blank}} * k * \frac{molarity\ of\ trolox}{weight\ of\ sample}$$

where k is the sample dilution factor, and AUC is the area below the fluorescence decay curve of the sample, blank, and Trolox, respectively. The area under the curve of normalized data was calculated using GraphPad Prism v. 6.0 (GraphPad Soft. Inc., La Jolla, San Diego, CA, USA).

The reducing ability of antioxidants toward the DPPH (2,2-diphenyl-1-picrylhydrazyl) radical was measured according to Brand-Williams et al. [29]. DPPH methanolic solution (684.7 μM) was adjusted with methanol until an absorbance of 0.550 ± 0.01 units at 517 nm (Genesys 20, Thermo Scientific, Waltham, MA, USA) was obtained. Samples (100 μL) were incubated with 1.45 mL of this DPPH solution for 60 min in the dark. A calibration curve was prepared from 1.95–150 μM Trolox ($r^2 = 0.99$). Results are expressed as μM Trolox equivalent per g of cocoa beans (dry matter basis).

2.8. Statistical Analysis

All the experimental measurements were repeated at least three times, and data were expressed as the mean ± standard deviation. Statistical analysis was done using Statistica 7.1 (Stat-Soft Inc., USA). The analysis of variance (ANOVA) and *F*-test were used to evaluate the influence of the factors and their interactions on the experimental designs. ANOVA one-way and Tukey's multiple range test was also conducted at a 5% level of significance. The kinetic constants in this study were determined from experimental data using non-linear regression employing Quasi-Newton method and least squares as custom loss function (0.0001 for both start value and initial step size). The response surface methodology consisting of full factorial central composite rotatable design with four replicates at the central point was conducted according to a completely randomized model. A second-order polynomial equation was used to fit the experimental data as follows:

$$Y = \beta_0 + \sum_{i=1}^{k} \beta_i X_i + \sum_{i=1}^{k} \beta_{ii} X_i^2 + \sum_{\substack{i=1 \\ i<j}}^{k-1} \sum_{j=2}^{k} \beta_{ij} X_i X_j$$

where Y is the predicted factor, β_0 is the value of the fitted response to the design, β_i, β_{ii}, and β_{ij} are the coefficients of linear, quadratic, and cross product terms, respectively.

The performance of full factorial central composite design method was measured by r and r^2. Experimental runs were also randomized to evaluate the concordance of experimental data and predicted values; therefore, the root mean squared error (RMSE) was calculated as follows:

$$RMSE = \sqrt[2]{\frac{\sum_{i=1}^{n}(y_i - \hat{y}_i)^2}{n}}$$

where y_i and $\hat{y}_i$ is the measured value and predicted value by the model, respectively. In addition, n is the number of the set data.

3. Results and Discussion

3.1. Effect of Drying Temperature, Particle Size and Non-Degreasing Process on the Concentration of Total Phenols

3.1.1. Drying Technology

Beans with reduced enzyme activity were manually chopped, and the total phenol content was evaluated after (a) oven-air drying at 50 °C and 70 °C, and (b) freeze-drying. The initial moisture content in a fresh unfermented bean was 40.1 wt.% and after drying process the final moisture content was between 1.4–3.2 wt.%. Table 1 summarizes the effect of drying process on TP content. Therefore, high temperature air drying led to a great extent of dehydration levels. Visual evidence showed that dried beans were slightly grey/purple color, which is characteristic of dried unfermented cocoa beans with high procyanidin content [12]. However, drying at 50 °C for 20 h resulted in TP content 10% lower than drying at 70 °C for 3 h ($p < 0.05$), which showed that polyphenols were more sensitive to drying time than temperature. No significant differences in TP amount was observed between freeze drying and oven-dried at 70 °C treatments, which was also reported by Kothe et al. [30], who found that when employing high temperatures and short times (100–160 °C for 30 min), different reactions can occur, such as configuration rearrangement of flavan-3-ols and epimerization, which could led to the formation of new flavan-3-ol products with higher phenol activity. Thereby, oven-air drying at 70 °C was chosen as drying technology for further experiments. Moreover, air drying is a more cost-effective solution (and faster) than freeze-drying.

3.1.2. Impact of Particle Size on Extraction Yield

After the PPO enzymatic inhibition and drying steps, the samples were milled at low temperature for a short time, thus avoiding the fat melting during the reduction of particle size. In an attempt to discriminate which sizes led to better performance on extraction yield, three distinct samples varying on particles size were evaluated as follow: $(S_1) > 0.42$ mm, (S_2) 0.42–0.18 mm, and $(S_3) < 0.18$ mm. The results showed that the best treatment was S_3 with a TP recovery of 50 mg GAE/g being 2× and 1.2× fold higher than S_1 and S_2, respectively. This study reinforces the idea that there is an inverse relationship between the particle size and the extraction yield. This behavior was also observed by Sun et al. [31], who found that extraction yields were 6 fold higher when particle size decreased from 2 to 0.074 mm.

Table 1. Effect of drying process on total polyphenol amount in cocoa beans.

Drying Method	Enthalpy of Sublimation (kJ/mol) *	Time (h)	Total Polyphenol Content (mgGAE/g)
Freeze Drying	51.00	48	43.99 ± 0.25 [a]
Oven-air (50 °C)	41.69	20	38.96 ± 1.47 [b]
Oven-air (70 °C)	40.84	3	43.20 ± 0.73 [a]

Mean value ± standard deviation ($n = 5$) with similar letters mean are not significantly different by ANOVA One-way Tukey ($p > 0.05$). * The thermophysical property data for the determination of enthalpy of sublimation of pure compounds (water) was calculated thorough the libraries of the NIST ThermoData Engine (TDE) version 10.

3.1.3. Conditions to Avoid the Degreasing Process

Non-defatted samples and defatted samples with a total fat content of 58.5% and 2.5%, respectively, were used for these experiments. In prior assays, results showed that TP extraction from the defatted samples was 60% higher than in non-defatted samples, which confirms that fat is a barrier for the diffusion of the bioactive compounds from the solid phase to the extractor solvent. Since one of the aims of the present work was to avoid the use of toxic organic solvents, ultrasonic treatment was selected as an alternative clean extraction technology. Thus, non-defatted cocoa samples were placed in a test tube and mixed with 50% ethanol (1/60 *w/v* ratio solute/solvent) and ultrasonicated (35 kHz, 4 °C). The effect of ultrasonic time (0–60 min) was tested as a function of TP amount. As can be seen in Figure 1, ultrasound increases the polyphenols extraction efficiency up to a 30 min. This is because the cavitation forces improving the polarity of the system; moreover, the bubbles in the liquid/solid extraction can explosively collapse and generate localized pressure causing plant tissue rupture [32], thus increasing the mass transfer rate. Figure 1 also shows that samples ultrasonicated for 30 min had no significant difference with the defatted control ($p < 0.05$). Thereby, the ultrasonic bath for 30 min was chosen as optimal value for further experiments.

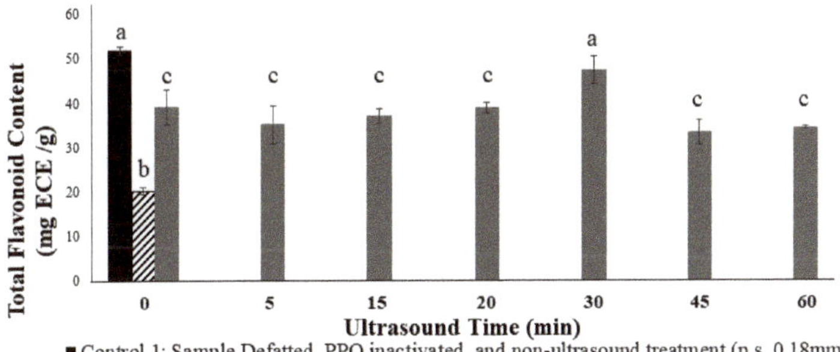

Figure 1. Impact of time of ultrasonic treatment on total polyphenolic amount. Means with different letters were significantly different by Tukey ($p < 0.05$).

3.1.4. Microscopy Analysis

Figure 2 shows SEM photographs confirming the mechanical effect of ultrasound on cell wall structure. As can be observed, non-treated samples had intact cell walls which were oval-shaped and not fractured, as well as appearing more solid and denser due to the cellular contents remaining embedded in the cell (Figure 2A). In general, the cocoa bean is formed of parenchyma cells (containing cocoa butter and proteins), which represent over 80% of the mass [33], and polyphenols and alkaloids, which are placed into the vacuoles [34]. Physical changes were observed in the structure of cocoa starch granules that become larger, smooth and fibrous, quite likely as produced by heat treatment (Figure 2B). This behavior could be the result of enzyme heat treatment, which opens up the cell and favors diffusion of bioactive compounds from the matrix into the extraction medium, as well as to the mobilization of proteins and polyphenols and redisposition of fat within the cell [34].

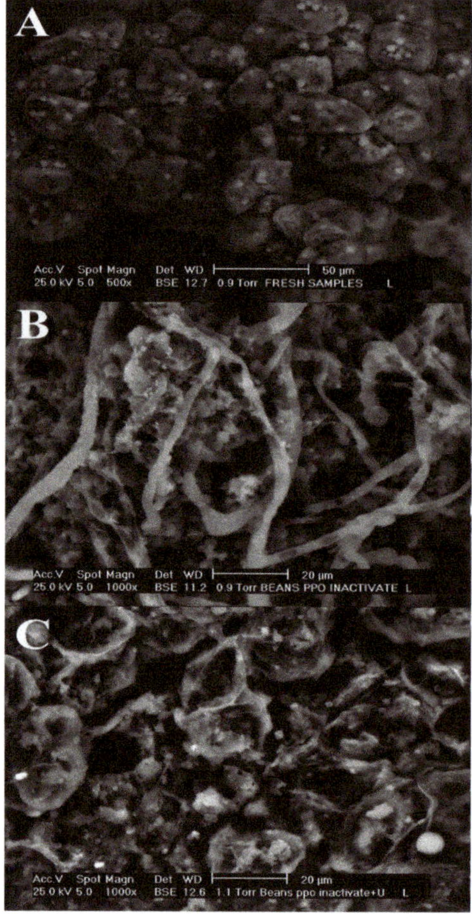

Figure 2. Microscopy images for the microstructure of (**A**) non-treated cocoa bean, (**B**) beans after PPO inhibition, and (**C**) beans after PPO inhibition and ultrasound treatment.

The morphological differences are more evident in Figure 2C. Smaller fragments were dispersed within the cell, and the microstructure was more porous, which could be the result of cell disruption and changes in the intercellular spaces. Hence, changes in the cell might increase the permeability and thus diffusion of polyphenols out of the cells. In accordance with our results, Rostagno et al. [35] found that disruption of tissue and cell walls are most efficient when heat treatment together with ultrasound are used, which resulted in a greater penetration of solvent into the sample matrix, increasing the contact surface between solid and liquid phase, and as a result, the solute quickly diffuses from the solid phase to the solvent. The observation of this study confirms that the enzymatic inhibition by heat treatment, milling process, and ultrasounds play a significant role in the change of the internal structure of cocoa beans, thus enhancing a high concentration of polyphenols while avoiding the defatting process.

3.2. Solid–Liquid Extraction of Polyphenols from Cocoa Beans

As mentioned, temperature, solute/solvent ratio, ethanol/water ratio, and pH were evaluated through a surface experimental design. As can be seen in Table 2, the concentration of ethanol had a strong impact on response factors, for instance, TP, TF and TF3 varied from 41.3–107.6 mg GAE/g, 15.2–86.0 mg ECE/g, and 21.7–59.8 mg ECE/g, respectively, throughout the ethanol concentration range (0–100%). Moreover, TP yield increased with increasing extraction temperature, i.e., extraction at 60 °C (25% ethanol, pH 5, 1/60 solvent/solute ratio) was 1.12 fold higher at 40 °C. Maximum values for TP (107.6 mg GAE/g), TF (86.0 mg ECE/g), and TF3 (59.8 mg ECE/g) were observed at different levels of factors, which indicate that combination of factors plays a key role in the extraction process.

Table 2. 2^4 full factorial central composite rotatable design and experimental results for total polyphenol, total flavonoids, and total flavan-3-ols recovery from cocoa beans.

T (°C)	SS (w/v)	EW (v/v)	pH	TP (mgGAE/g)	TF (mgECE/g)	TF3 (mgECE/g)
60	1/30	25	5	78.22 ± 1.17	62.30 ± 1.75	35.07 ± 0.14
50	1/24	50	4	94.15 ± 0.50	79.64 ± 1.88	48.62 ± 0.59
50	1/40	50	2	98.86 ± 0.52	55.87 ± 2.60	39.46 ± 0.25
60	1/60	75	3	103.57 ± 1.25	57.48 ± 0.39	43.78 ± 0.90
40	1/30	75	3	92.23 ± 1.79	77.30 ±4.10	42.89 ± 0.27
60	1/60	25	3	91.59 ± 1.30	52.36 ± 0.39	40.74 ± 0.98
50	1/40	50	2	98.17 ± 1.10	53.84 ± 2.34	38.01 ± 0.13
40	1/30	75	3	95.81 ± 0.07	68.77 ± 3.51	40.75 ± 0.24
50	1/40	50	6	98.85 ± 0.65	66.45 ± 2.08	59.28 ± 0.98
50	1/40	50	4	95.96 ± 0.82	67.97 ± 4.16	51.26 ± 0.09
50	1/120	50	4	107.59 ± 0.52	53.74 ±0.77	55.21 ± 1.39
50	1/40	50	6	97.53 ± 0.26	64.31 ± 0.26	59.78 ± 1.54
60	1/30	75	3	97.90 ± 0.93	73.11 ± 2.35	47.46 ± 0.04
50	1/40	50	4	95.52 ± 0.61	63.01 ± 3.38	47.38 ± 0.93
40	1/30	75	5	81.23 ± 0.23	76.18 ± 4.30	50.28 ± 1.40
40	1/30	25	5	76.97 ± 1.29	61.21 ± 0.96	42.77 ± 0.28
40	1/60	75	5	91.56 ± 0.19	49.34 ± 2.72	42.19 ± 0.63
60	1/30	25	3	82.64 ± 0.13	59.53 ± 0.59	39.84 ± 0.07
60	1/60	75	5	89.26 ± 0.45	57.78 ± 0.39	58.68 ± 1.12
50	1/40	100	4	46.79 ± 0.44	15.18 ± 1.30	22.72 ± 0.37
40	1/60	25	3	84.58 ± 0.19	50.93 ± 3.12	38.83 ± 0.99
40	1/30	25	3	76.06 ± 0.69	55.74 ± 2.91	43.87 ± 0.21
40	1/60	25	3	84.45 ± 0.99	51.83 ± 3.52	40.37 ± 1.28
30	1/40	50	4	88.59 ± 0.13	74.73 ± 1.56	42.52 ± 0.14
70	1/40	50	4	100.04 ± 0.69	79.73 ± 4.14	58.04 ± 0.51
60	1/60	25	3	82.98 ± 2.23	55.26 ± 1.16	39.85 ± 0.85
40	1/30	25	5	78.95 ± 1.03	59.04 ± 3.69	46.46 ± 0.21
40	1/60	75	3	102.28 ± 2.25	46.02 ± 3.91	29.49 ± 0.61
60	1/30	25	5	81.21 ± 0.45	64.10 ± 3.88	38.59 ± 0.36
50	1/120	50	4	104.96 ± 0.91	55.91 ± 1.56	55.57 ± 0.56
60	1/30	75	5	83.48 ± 0.58	74.43 ± 4.27	55.98 ± 0.52
50	1/40	0	4	43.45 ± 1.22	22.40 ± 1.56	21.73 ± 0.09
40	1/60	25	5	76.43 ± 1.17	56.35 ± 2.34	41.86 ± 0.42
50	1/40	0	4	41.32 ± 0.22	17.09 ± 0.26	25.81 ± 0.65
30	1/40	50	4	89.97 ± 0.91	77.91 ± 2.60	41.31 ± 0.23
40	1/30	25	3	79.12 ± 0.67	58.51 ± 0.39	39.37 ± 0.25

Table 2. *Cont.*

T (°C)	SS (*w/v*)	EW (*v/v*)	pH	TP (mgGAE/g)	TF (mgECE/g)	TF3 (mgECE/g)
60	1/30	75	5	84.29 ± 0.39	75.47 ± 4.26	52.63 ± 1.39
60	1/30	25	3	82.19 ± 0.39	59.47 ± 3.50	36.21 ± 0.39
60	1/60	75	5	89.01 ± 1.36	57.96 ± 1.16	54.35 ± 1.74
50	1/40	50	4	90.36 ± 0.48	69.19 ± 4.17	44.92 ± 0.84
60	1/60	25	5	85.11 ± 0.33	59.55 ± 1.95	44.08 ± 0.35
50	1/24	50	4	100.27 ± 0.05	86.05 ± 4.53	53.69 ± 0.25
70	1/40	50	4	97.97 ± 0.26	82.05 ± 2.08	58.80 ± 0.84
60	1/60	75	3	95.49 ± 2.51	53.52 ± 0.78	46.89 ± 1.50
60	1/30	75	3	98.69 ± 0.57	72.95 ± 3.50	49.99 ± 1.22
50	1/40	50	4	93.54 ± 1.13	64.19 ± 1.30	45.78 ± 0.77
40	1/30	75	5	81.22 ± 0.49	70.55 ± 2.15	47.68 ± 0.77
50	1/40	50	4	90.52 ± 0.30	65.81 ± 4.16	50.50 ± 0.79
50	1/40	50	4	91.27 ± 0.17	64.09 ± 3.90	50.16 ± 0.19
40	1/60	25	5	77.75 ± 0.91	51.53 ± 0.39	41.35 ± 1.01
40	1/60	75	3	99.96 ± 1.39	45.00 ± 1.55	32.91 ± 1.12
60	1/60	25	5	87.44 ± 1.25	56.65 ± 2.73	41.67 ± 0.85
50	1/40	100	4	47.40 ± 0.35	15.19 ± 0.78	23.69 ± 0.28
40	1/60	75	5	92.33 ± 0.32	48.27 ± .50	46.59 ± 0.14
Control by S-L extraction (Yield = 14.9%)				49.35 ± 2.06 [a]	35.71 ± 0.19 [a]	26.41 ± 1.88 [a]
Optimum by S-L Extract. (Yield = 16.8%)				122.34 ± 2.35 [b]	88.86 ± 0.78 [b]	62.57 ± 3.37 [b]

where TP is total polyphenol TF is total flavonoid and TF3 is total flavan-3-ol. S-L is the solid–liquid extraction. Means (n = 3) within a column (comparison of same bioactive compound family) with different letters were significantly different by ANOVA One-way Tukey ($p < 0.05$). For more information see the methodology section.

Results were fitted to a quadratic model (Equations (S1)–(S3)) and the 3D surface graphs for dependent variables were plotted (Figure 3). Thus, analysis of variance summarized the observations and differences among the studied factors (Table S1). ANOVA showed that the recovery of bioactive compounds was mainly dependent on the linear effect of temperature (T) and pH, linear and quadratic effect of ethanol/water (EW) ratio, and solute/solvent (SS) ratio. Besides, the interactions of T × EW, and EW × pH for total polyphenol content; SS × T, and SS × EW for total flavonoid concentration; and T × SS, T × EW, SS × pH, and EW × pH for total flavan-3-ol amount were also significant ($p < 0.05$). ANOVA analysis also showed that the selected quadratic model adequately represented the extraction process. Therefore, the model has a good coefficient of multiple determination of $r^2 = 0.957$, $r_{adj} = 0.941$; $r^2 = 0.954$, $r_{adj} = 0.937$; and $r^2 = 0.905$, $r_{adj} = 0.872$ for total polyphenol, flavonoid and flavan-3-ols content, respectively. Overall, higher accordance regression fit values mean that the model explained most of the variability in the responses.

Thereby, the maximum experimental conditions were enhanced with 50/50 *v/v* ethanol/water ratio, 1/120 *w/v* solute/solvent ratio, pH 6 at 70 °C for a predicted recovery of 117.87 ± 16.68 mgGAE/g, 85.22 ± 18.51 mgEC/g, and 76.86 ± 15.98 mgECE/g, which were in agreement with the experimental data with 122.34 ± 2.35 mg GAE/g, 88.87 ± 0.78 mgECE/g, and 62.57 ± 3.37 mgECE/g, for TP, TF, and TF3, respectively.

Effect of Independent Factors on the Recovery of Total Polyphenols and Total Flavonoids

As can be seen in Figure 3, the significant factors evaluated had a parabolic behavior on the extraction of TP, TF, and TF3. For instance, it was confirmed that temperature promoted faster diffusion rate, better mass transfer, reduce viscosity and surface tension (Figure 3A,C,E), which is in agreement with Cacace and Mazza [36]. Indeed, this factor had a linear effect until it reached a maximum, after which the extraction yield decreased. This phenomenon is explained by the softening of plant tissue at high temperature, while at the same time, it weakens phenol-protein and phenol-polysaccharide interactions [37]. Moreover, results showed that higher diffusion rate is enhanced at lower ratio solute/solvent, which is confirmed by the significance of linear and quadratic effect for both total phenol and total flavonoid content. Also, Fick´s second law of diffusion predicts this phenomenon, that is, a final equilibrium between the concentration of solute in the solid matrix and in the bulk solution after a certain time [37].

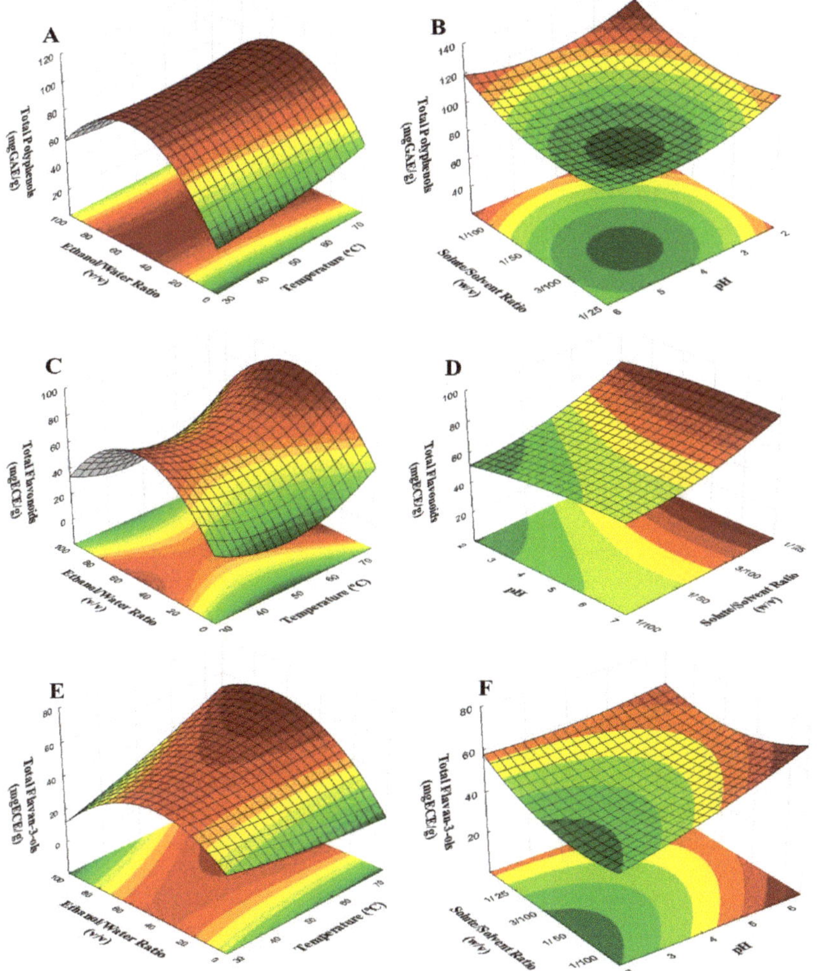

Figure 3. Surface response as function for Temperature vs. Ethanol concentration, and pH vs. solute concentration for the recovery of total polyphenols (**A,B**), total flavonoids (**C,D**) and total flavan-3-ols (**E,F**).

The influence of ethanol concentration (Figure 3A,C,E) suggested that the extract contains polyphenols with different polarities, which also confirms the principle "like dissolves like" The lowest concentration of water or ethanol were not efficient for the extraction of flavonoids from cocoa beans. The maximum yield was enhanced at 50% ethanolic aqueous solution. Previous works are consistent with our results, reporting a maximum extraction yield of polyphenols with an ethanol/water ratio of 40–50% [18,38], which is because the concentration of ethanol influences the velocity-transfer kinetic and the thermodynamics of the process [38].

In addition, the interaction of pH with the solvent (Figure 3B,D,F) plays an essential role in controlling adsorption/desorption rate (i.e., changing the surface charge of adsorbent, the degree of ionization and speciation of adsorbate during adsorption) [39], and an impact on the pKa value for the polyphenols. In fact, lower solvent acidity allowed the highest recovery of flavonoids, because acid could hydrolyze the cell wall and thus facilitates the diffusion of the phenolic compounds. However, at higher concentrations of ethanol, a reduction on the extraction yield was observed. Similar results show that the amount of acid in the extraction solvent increases concentration by increasing the stability of the phenolics during extraction and increasing their dissolution [37].

In addition, significant differences ($p < 0.05$) among control samples and sample at optimum conditions were found; thereby, TP, TF, and TF3 were 59.7, 59.8, and 57.8% higher, respectively, when employing the optimized S-L extraction method (Table 2).

3.3. Extraction Kinetics Parameters

Given the maximum conditions for the recovery of polyphenols, different extractions were carried out to evaluate the equilibrium time and kinetic parameters as a function of TP, TF, and TF3 content.

The study of equilibrium time plays an essential role in economizing the energy and cost of the industrial process [37], thus improving the accuracy of the procedure and the quality of the final product. Table 3 summarizes the kinetic parameters, RMSE and the accordance of the model (r) for all the mathematical models selected, which were previously used to model the solid-to-liquid extraction of bioactive compounds. Overall, TP, TF, and TF3 content had the lowest r and highest RMSE for those equations that consider the extraction is occurring in one continuous step, i.e., Equations (1), (2), (4), and (5). Goodness of fit of the model and lower standard deviation of residuals were obtained for the two models that represent the recovery of polyphenols on two different rates (sorption/desorption), for instance, Equation (3) and Equation (6). These equations represent the kinetic process of a liquid/solid system based on the solid capacity. As can be seen in Figure 4, the extraction curve shape had a faster extraction rate followed by a slower extraction rate and asymptotically approached the equilibrium concentration. Generally, the temperature enhances the solubility of polyphenols in the solvent [40], but the rate of extraction yield decreased progressively due to the exposure to both higher temperature and time extraction, which confirms that polyphenols are thermosensitive compounds [41].

Results suggested that the Peleg model (Equation (3)) proved to be most suitable to model the solid–liquid extraction kinetics for the dependent variables ($r \geq 0.98$ and RMSE ≤ 0.71). In fact, k_1 and k_2 represent the extraction rate constant and constant of extraction extent, respectively [42]. Table 3 shows that k_1 was similar in all the cases, but k_2 increased by TF3 ~ TF >> TP, which is related to its maximum equilibrium concentration ($C_{t \to \infty}$). Thereby, the equilibrium extraction time were 45 min, 39 min and 34 min for TP, TF, and TF3 content, respectively. Comparison of these data with those of previous authors shows that our optimized extraction method was 2.7×, 4.4×, 6.0×, and 2.7× faster than the polyphenol recovery of cocoa beans [43], grape seed [42], and mango kernel [44], respectively.

In addition, a large-scale extraction process was also assayed. The results showed that recovery from 0.05 to 5 L did not significantly impact extraction yield ($p < 0.05$), but the extraction at 10 L was 25.7 ± 0.9% lower for TP content while was 23.0 ± 6.9% and 5.3 ± 2.6% higher for TF and TF3 amount, respectively. These differences could be due to diffusional changes as a result of the scaling up the process.

In general, this work confirmed the importance of studying the impact of extraction parameters on both secondary plant metabolites, thereby increasing the yield of extraction. In terms of going beyond the highlights, the optimized ultrasound-assisted solid–liquid extraction not only allowed a high concentration of both total polyphenols and flavonoids but also employed food-grade solvents, reduced the number of stages (i.e., avoiding the degreasing), the extraction time (<45 min), and therefore energy consumption. Thereby, the extraction process could be suitable for large-scale applications. For example, cocoa polyphenol extract can be used to enrich products in very high demand in the food and cosmetic industries.

Table 3. Kinetic models used for the fitting for total polyphenol, total flavonoids, and total flavan-3-ols content from cocoa beans.

Model		Parameters for TP	Parameters for TF	Parameters for TF3	Ref.
nth order (1)	$C(t) = kt^n$	$k = 55.64$ $n = 0.16$	$k = 46.63$ $n = 0.13$	$k = 37.03$ $n = 0.11$	[26]
	r	0.93	0.90	0.90	
	RMSE	8.23	5.46	4.19	
Page (2)	$C(t) = e^{kt^n}$	$k = 4.06$ $n = 0.03$	$k = 3.87$ $n = 0.03$	$k = 3.64$ $n = 0.02$	[23]
	r	0.93	0.89	0.89	
	RMSE	8.52	5.55	4.22	
Peleg (3)	$C(t) = \dfrac{t}{k_1 + k_2 t}$	$k_1 = 0.03$ $k_2 = 8.1 \times 10^{-3}$	$k_1 = 0.03$ $k_2 = 0.01$	$k_1 = 0.03$ $k_2 = 0.01$	[24]
	r	0.98	0.99	0.99	
	RMSE	0.71	0.62	0.67	
Weibull-type (4)	$C(t) = C_0 e^{kt^n}$	$C_0 = 1.7 \times 10^{-8}$ $k = 21.92$ $n = 6.9 \times 10^{-3}$	$C_0 = 2.8 \times 10^{-7}$ $k = 13.17$ $n = 6.4 \times 10^{-3}$	$C_0 = 2.0 \times 10^{-7}$ $k = 19.03$ $n = 5.8 \times 10^{-3}$	[22]
	r	0.93	0.89	0.90	
	RMSE	8.33	5.51	4.32	
Mincher and Minkov (5)	$C(t) = A - Be^{-kt}$	$A = 96.94$ $B = 96.94$ $k = 289.60$	$A = 72.29$ $B = 72.29$ $k = 103.69$	$A = 55.07$ $B = 55.07$ $k = 117.03$	[26]
	r	0.64	0.68	0.71	
	RMSE	22.06	12.47	8.35	
Pseudo first order (6)	$C(t) = C_\infty - \dfrac{C_\infty}{e^{kt+a}}$	$C_\infty = 122.34$ $k = 0.11$ $a = 0.124$	$C_\infty = 88.87$ $k = 0.18$ $a = 0.024$	$C_\infty = 62.57$ $k = 0.25$ $a = 0.020$	[25]
	r	0.98	0.98	0.99	
	RMSE	6.49	5.59	2.29	

where TP is total polyphenol (mgGAE/g) TF is total flavonoid (mgECE/g), and TF3 is total flavan-3-ol (mgECE/g).

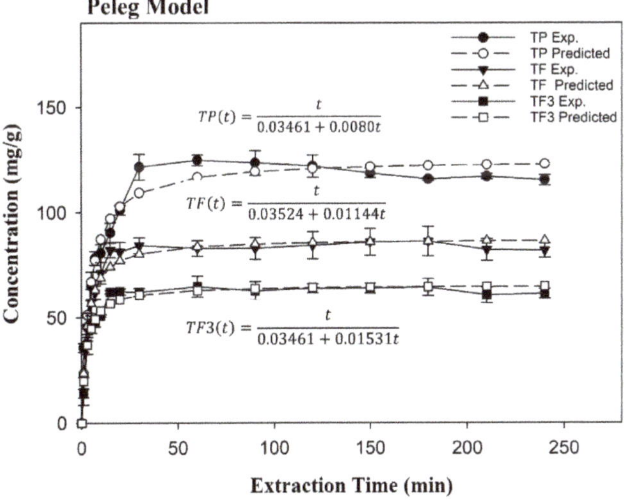

Figure 4. Experimental and calculated extraction curves for a total polyphenol (TP), total flavonoids (TF), and flavan-3-ols content. E (TF3). Extraction rate constant based on Peleg model and solid–liquid extraction at optimal conditions.

3.4. Chromatographic and Antioxidant Analysis

Identification of alkaloids, catechins and procyanidins in cocoa, were achieved by HPLC-DAD-ESI-MS/MS. The chromatographic method allowed not only the detection of main alkaloids but also the procyanidins with different degree of polymerization up to 7 (Figure 5). As can be seen in Table 4, the cocoa extract obtained at maximum conditions contained 65% (*w/w*) procyanidins followed by methylxanthines (20% *w/w*) and catechins (15% *w/w*), which is in agreement with previous authors [45,46]. Theobromine/caffeine ratio was 2.93, which can be used to classify hybrid genotypes [47]. Therefore, it was confirmed that the cocoa sample was of the Trinitarian variety. (−)-Epicatechin (7.30 mg/g) was the main catechin, being 11.8 fold higher than (+)-catechin, which was also confirmed by Romero et al. [48]. The major procyanidin in cocoa polyphenol extract was the Trimer C1 (11.9 mg/g) being 2.9×, 1.3×, 1.7×, 6.6×-fold higher than the dimer, tetramer, pentamer, and hexamer, respectively.

Our results are in agreement with previous works that reported that the main alkaloid and flavanol are the theobromine and (−)-epicatechin, but their concentrations vary considerably depending on the methodology of extraction and cocoa bean variety [49]. Theobromine and caffeine concentrations were in agreement with previous reports [8,15], but epicatechin and catechin amounts were 1.5× and 1.9×, and 2.0× and 2.3× fold higher than those reported by Kothe et al. [30] and Carrillo et al. [15], respectively. In general, procyanidin content was much higher than previous authors, i.e., dimer B2 was 2.0× and 1.5× -fold greater than procyanidin B2 previously reported by Kothe et al. [30] and Tomas-Barberán et al. [43], respectively.

In addition, the identification of oligomeric procyanidins was carried out by mass spectrometry in comparison with commercial standards and published literature. For instance, dimer B2 (B-type, EC-4β→8-EC-4β), trimer C1 (B-type, EC-4β→8-EC-4β→8-EC), tetramer, pentamer, hexamer and heptamer with a molecular ion [M − H]⁻ *m/z* 577, 865, 1153, 1441, 1729, 2019, respectively were identified. Characterization of larger polymers was not possible due to their low concentration, low ionization, peak signal dispersion, the formation of multiple ions, and limitations of the ion trap MS analyzer. Fragment patterns also suggested that various fragmentation mechanisms are involved in ESI such as quinone methide (QM), retro-Diels-Alder (RDA), as well as heterocyclic ring fission (HRF), which could take place on the extension unit or the terminal unit of the molecule [50]. For instance, the loss of a fragment with *m/z* 152, 170 and additional loss of water corresponding to RDA fission and loss of 288 Da corresponds to QM cleavage (Figure 5).

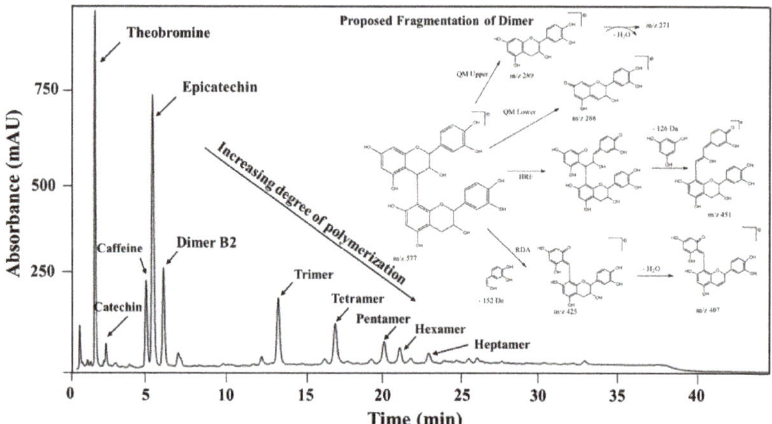

Figure 5. Chromatogram of cocoa extract and hypothetical electrospray ionization (negative mode) fragmentation pathway for procyanidin dimer. RDA, retro-Diels-Alder fission; QM, quinone methide cleavage; HRF, heterocyclic ring fission.

Table 4. Methylxanthine and procyanidin concentration, and characterization of cocoa beans using a HPLC-DAD-ESI-MS/MS method.

| Compound | Reverse Phase | | Ionization | Predicted Mass (mau) | Observed Mass (mau) | Error (mau) | HPLC-ESI-MS^n |
	Retention Time (min)	Concentration* (ppm)					MS^2 Fragment (*m/z*)
Theobromine	1.42	7.78 ± 0.01	[M + H]$^+$	181.07	181.8	0.73	137.5, 110.5
Caffeine	5.01	2.65 ± 0.02	[M + H]$^+$	195.08	195.5	0.42	158.4, 138.7
Catechin	2.16	0.62 ± 0.01	[M − H]$^+$	291.08	292.3	1.22	273.3, 165.3, 139.4, 123.7
Epicatechin	5.46	7.30 ± 0.10	[M − H]$^+$	291.08	292.3	1.22	273.3, 165.3, 139.3, 123.6
Dimer B2	6.22	4.06 ± 0.03	[M − H]$^-$	577.14	577.4	0.26	451.2, 425.1, 289.1, 271.1
Trimer C1	12.91	11.99 ± 0.25	[M − H]$^-$	865.19	865.4	0.21	695.2, 577.2, 451.0, 289.0
Tetramer D	16.81	9.33 ± 0.40	[M − H]$^-$	1153.26	1153.6	0.34	1027.3, 865.3, 739.2, 577.1
Pentamer	19.91	6.81 ± 0.52	[M − H]$^-$	1441.33	1441.3	0.03	1153.3, 865.2, 691.6, 574.3
Hexamer	20.85	1.81 ± 0.01	[M − H]$^-$	1729.38	1729.3	0.08	1534.0, 1153.3, 865.2, 574.2
Heptamer	22.51	ND	[M − H]$^-$	2017.45	2019.3	1.85	1153.4, 995.3, 851.3, 574.3

Data expressed as means of triplicate experiments. * Concentration expressed as mg polyphenol per g cocoa beans (dry weight basis).

Among the antioxidant assays, our results showed that the cocoa extract had an ORAC of 1149.85 ± 25.1 µM Trolox eq/g. These values are higher than those reported by Hurst et al. [51] and Carrillo et al. [15], with TP in the range 58.0–61.7 and 45.3–70.0 mg GAE/g, and ORAC in the range 797.0–947.0 and 387.3–618.1 µM Trolox eq/g, respectively. Moreover, DPPH radical scavenging activity was 120.6 ± 0.5 µM Trolox eq/g (equal to 0.72 µM Trolox eq/mg cocoa extract), which was 2.4 fold higher than previously reported by Summa et al. [52]. These differences could be associated with the improved methodology of extraction at optimal conditions, thus enhancing both high composition and concentration of catechins (7.92 mg) and procyanidins (34.0 mg) with a DP ≥ 7.

4. Conclusions

An effective ultrasound-assisted solid–liquid process for extracting polyphenols from cocoa beans was optimized using an experimental design. The 2^4 factorial design showed that all parameters studied were significant factors in affecting the polyphenolic content as well as enabling us to determine the optimal values for the extraction process. This optimization showed that the best conditions to obtain high polyphenol yield were 50% of ethanol, solid/solvent ratio of 1:120 *w/v*, pH 6 at 70 °C for a maximum equilibrium time of 45 min. Operating conditions to avoid degreasing and freeze-drying steps were also established, thus leading to a more cost-effective strategy. Overall, the process extraction allowed to increase on 59.7% and 12.8% of cocoa polyphenols amount and extraction yield, respectively. The extract rich in polyphenols could replace the synthetic antioxidants and could be used in the food and cosmetic industries. Our results suggest that ultrasound-assisted solid–liquid extraction is a suitable method for the recovery of cocoa polyphenols and could be used for the scaleup procedure for further research.

Supplementary Materials: The following are available online at http://www.mdpi.com/2076-3921/9/5/364/s1, Table S1: ANOVA for TP, TF, and TF3 through 24 surface design + central points+ start points.

Author Contributions: Formal analysis, E.I., E.A.D. and L.J.L.-G.; Investigation, S.T.-U.; Methodology, S.T.-U.; Supervision, E.I., E.A.D., A.R.V.-J. and L.J.L.-G.; Writing—original draft, S.T.-U.; Writing—review & editing, L.J.L.-G. All authors have read and agreed to the published version of the manuscript.

Funding: This research was funded by Universidad Industrial de Santander, grant number 2511. In addition, to Department of Science, Technology and Innovation (COLCIENCIAS-Colombia), grant number 567 of 2012.

Acknowledgments: Said Toro wants to thank Colciencias (Colombia) for my financial scholarship, University of Massachusetts, Foodomics Laboratory. Also, the authors want to thank Vicerrectoría de Investigación y Extensión from Universidad Industrial de Santander for their financial support that made this work possible.

Conflicts of Interest: The authors declare no conflict of interest.

References

1. Pedan, V.; Fischer, N.; Rohn, S. An online NP-HPLC-DPPH method for the determination of the antioxidant activity of condensed polyphenols in cocoa. *Food Res. Int.* **2016**, *89*, 890–900. [CrossRef]
2. Batista, N.N.; de Andrade, D.P.; Ramos, C.L.; Dias, D.R.; Schwan, R.F. Antioxidant capacity of cocoa beans and chocolate assessed by FTIR. *Food Res. Int.* **2016**, *90*, 313–319. [CrossRef] [PubMed]
3. Toro-Uribe, S.; Montero, L.; López-Giraldo, L.; Ibáñez, E.; Herrero, M. Characterization of secondary metabolites from green cocoa beans using focusing-modulated comprehensive two-dimensional liquid chromatography coupled to tandem mass spectrometry. *Anal. Chim. Acta* **2018**, *1036*, 204–213. [CrossRef] [PubMed]
4. Pérez-Jiménez, J.; Neveu, V.; Vos, F.; Scalbert, A. Identification of the 100 richest dietary sources of polyphenols: An application of the Phenol-Explorer database. *Eur. J. Clin. Nutr.* **2010**, *64*, S112–S120. [CrossRef] [PubMed]
5. Presidencia de la República de Colombia Decreto Ley 896 de 2017 "Programa Nacional Integral de Sustitución de Cultivos de uso ilícito PNIS". Available online: http://www.indepaz.org.co/wp-content/uploads/2017/05/Decreto-896-del-29-de-Mayo-de-2017-1.pdf (accessed on 3 August 2018).
6. Quiroz-Reyes, C.N.; Aguilar-Mendez, M.A.; Ramírez-Ortíz, M.E.; Ronquillo-De Jesus, E. Comparative study of ultrasound and maceration techniques for the extraction of polyphenols from cocoa beans (*Theobroma cacao* L.). *Rev. Mex. Ing. Química* **2013**, *12*, 11–18.
7. Routray, W.; Orsat, V. Microwave-Assisted Extraction of Flavonoids: A Review. *Food Bioprocess Technol.* **2012**, *5*, 409–424. [CrossRef]
8. Okiyama, D.C.G.; Soares, I.D.; Cuevas, M.S.; Crevelin, E.J.; Moraes, L.A.B.; Melo, M.P.; Oliveira, A.L.; Rodrigues, C.E.C. Pressurized liquid extraction of flavanols and alkaloids from cocoa bean shell using ethanol as solvent. *Food Res. Int.* **2018**, *114*, 20–29. [CrossRef]
9. Servent, A.; Boulanger, R.; Davrieux, F.; Pinot, M.N.; Tardan, E.; Forestier-Chiron, N.; Hue, C. Assessment of cocoa (*Theobroma cacao* L.) butter content and composition throughout fermentations. *Food Res. Int.* **2018**, *107*, 675–682. [CrossRef]
10. Rodríguez-Carrasco, Y.; Gaspari, A.; Graziani, G.; Sandini, A.; Ritieni, A. Fast analysis of polyphenols and alkaloids in cocoa-based products by ultra-high performance liquid chromatography and Orbitrap high resolution mass spectrometry (UHPLC-Q-Orbitrap-MS/MS). *Food Res. Int.* **2018**, *111*, 229–236. [CrossRef]
11. Żyżelewicz, D.; Budryn, G.; Oracz, J.; Antolak, H.; Kręgiel, D.; Kaczmarska, M. The effect on bioactive components and characteristics of chocolate by functionalization with raw cocoa beans. *Food Res. Int.* **2018**, *113*, 234–244. [CrossRef]
12. Schinella, G.; Mosca, S.; Cienfuegos-Jovellanos, E.; Pasamar, M.Á.; Muguerza, B.; Ramón, D.; Ríos, J.L. Antioxidant properties of polyphenol-rich cocoa products industrially processed. *Food Res. Int.* **2010**, *43*, 1614–1623. [CrossRef]
13. Pons-Andreu, J.-V.; Cienfuegos-Jovellanos, E.; Ibarra, A. Process for Producing Cocoa Polyphenol Concentrate. U.S. Patent 0193629 A1, 14 August 2008.
14. Pinelo, M.; Sineiro, J.; Núñez, M.J. Mass transfer during continuous solid-liquid extraction of antioxidants from grape byproducts. *J. Food Eng.* **2006**, *77*, 57–63. [CrossRef]
15. Carrillo, L.C.; Londoño-Londoño, J.; Gil, A. Comparison of polyphenol, methylxanthines and antioxidant activity in Theobroma cacao beans from different cocoa-growing areas in Colombia. *Food Res. Int.* **2014**, *60*, 273–280. [CrossRef]
16. Ioannone, F.; Di Mattia, C.D.; De Gregorio, M.; Sergi, M.; Serafini, M.; Sacchetti, G. Flavanols, proanthocyanidins and antioxidant activity changes during cocoa (*Theobroma cacao* L.) roasting as affected by temperature and time of processing. *Food Chem.* **2015**, *174*, 256–262. [CrossRef]
17. Patras, M.A.; Milev, B.P.; Vrancken, G.; Kuhnert, N. Identification of novel cocoa flavonoids from raw fermented cocoa beans by HPLC-MSn. *Food Res. Int.* **2014**, *63*, 353–359. [CrossRef]

18. Chew, K.K.; Khoo, M.Z.; Ng, S.Y.; Thoo, Y.Y.; Aida, W.M.W.; Ho, C.W. Effect of ethanol concentration, extraction time and extraction temperature on the recovery of phenolic compounds and antioxidant capacity of Orthosiphon stamineus extracts. *Int. Food Res. J.* **2011**, *18*, 1427–1435.

19. Singleton, V.L.; Orthofer, R.; Lamuela-Raventós, R.M. Analysis of total phenols and other oxidation substrates and antioxidants by means of folin-ciocalteu reagent. *Methods Enzymol.* **1998**, *299*, 152–178.

20. Zhishen, J.; Mengcheng, T.; Jianming, W. The determination of flavonoid contents in mulberry and their scavenging effects on superoxide radicals. *Food Chem.* **1999**, *64*, 555–559. [CrossRef]

21. Sun, B.; Ricardo-da-Silva, J.M.; Spranger, I. Critical Factors of Vanillin Assay for Catechins and Proanthocyanidins. *J. Agric. Food Chem.* **1998**, *46*, 4267–4274. [CrossRef]

22. Amendola, D.; De Faveri, D.M.; Spigno, G. Grape marc phenolics: Extraction kinetics, quality and stability of extracts. *J. Food Eng.* **2010**, *97*, 384–392. [CrossRef]

23. Doymaz, İ.; İsmail, O. Drying characteristics of sweet cherry. *Food Bioprod. Process.* **2011**, *89*, 31–38. [CrossRef]

24. Peleg, M. An Empirical Model for the Description of Moisture Sorption Curves. *J. Food Sci.* **1988**, *53*, 1216–1217. [CrossRef]

25. Spiro, M.; Jago, D.S. Kinetics and equilibria of tea infu- sion. Part 3. Rotating disc experiments interpreted by a steady state model. *J. Chem. Soc. Faraday Trans. 1* **1982**, *78*, 295–305. [CrossRef]

26. Sant'Anna, V.; Brandelli, A.; Marczak, L.D.F.; Tessaro, I.C. Kinetic modeling of total polyphenol extraction from grape marc and characterization of the extracts. *Sep. Purif. Technol.* **2012**, *100*, 82–87. [CrossRef]

27. Toro-Uribe, S.; López-Giraldo, L.J.; Decker, E.A. Relationship between the physiochemical properties of cocoa procyanidins and their ability to inhibit lipid oxidation in liposomes. *J. Agric. Food Chem.* **2018**, *66*, 4490–4502. [CrossRef] [PubMed]

28. Huang, D.; Ou, B.; Hampsch-Woodill, M.; Flanagan, J.A.; Prior, R.L. High-throughput assay of oxygen radical absorbance capacity (ORAC) using a multichannel liquid handling system coupled with a microplate fluorescence reader in 96-well format. *J. Agric. Food Chem.* **2002**, *50*, 4437–4444. [CrossRef]

29. Brand-Williams, W.; Cuvelier, M.E.; Berset, C. Use of a free radical method to evaluate antioxidant activity. *LWT Food Sci. Technol.* **1995**, *28*, 25–30. [CrossRef]

30. Kothe, L.; Zimmermann, B.F.; Galensa, R. Temperature influences epimerization and composition of flavanol monomers, dimers and trimers during cocoa bean roasting. *Food Chem.* **2013**, *141*, 3656–3663. [CrossRef]

31. Sun, Y.; Liu, D.; Chen, J.; Ye, X.; Yu, D. Effects of different factors of ultrasound treatment on the extraction yield of the all-trans-B-carotene from citrus peels. *Ultrason. Sonochem.* **2011**, *18*, 243–249. [CrossRef]

32. Goula, A.M. Ultrasound-assisted extraction of pomegranate seed oil—Kinetic modeling. *J. Food Eng.* **2013**, *117*, 492–498. [CrossRef]

33. Hoskin, J.M.; Dimick, P.S.; Daniels, R.R. Scanning Electron Microscopy of the Theobroma cacao Seed. *J. Food Sci.* **1980**, *45*, 1538–1540. [CrossRef]

34. Lopez, A.S.; Dimick, P.S.; Walsh, R.M. Scanning Electron Microscopy Studies of the Cellular Changes in Raw, Fermented and Dried Cocoa. *Food Struct.* **1987**, *6*, 9–16.

35. Rostagno, M.A.; Palma, M.; Barroso, C.G. Ultrasound-assisted extraction of soy isoflavones. *J. Chromatogr. A* **2003**, *1012*, 119–128. [CrossRef]

36. Cacace, J.E.; Mazza, G. Mass transfer process during extraction of phenolic compounds from milled berries. *J. Food Eng.* **2003**, *59*, 379–389. [CrossRef]

37. Mokrani, A.; Madani, K. Effect of solvent, time and temperature on the extraction of phenolic compounds and antioxidant capacity of peach (*Prunus persica* L.) fruit. *Sep. Purif. Technol.* **2016**, *162*, 68–76. [CrossRef]

38. Rakotondramasy-Rabesiaka, L.; Havet, J. Estimation of effective diffusion and transfer rate during the protopine extraction process from *Fumaria officinalis* L. *Sep. Purif. Technol.* **2010**, *76*, 126–131. [CrossRef]

39. Datta, C.; Dutta, A.; Dutta, D.; Chaudhuri, S. Adsorption of polyphenols from ginger rhizomes on an anion exchange resin Amberlite IR-400—Study on effect of pH and temperature. *Procedia Food Sci.* **2011**, *1*, 893–899. [CrossRef]

40. Galvan D'Alessandro, L.; Kriaa, K.; Nikov, I.; Dimitrov, K. Ultrasound assisted extraction of polyphenols from black chokeberry. *Sep. Purif. Technol.* **2012**, *93*, 42–47. [CrossRef]

41. Othman, S.N.S.; Mustapa, A.N.; Ku Hamid, K.H. Extraction of polyphenols from Clinacanthus nutans Lindau (*C. nutans*) by vacuum solvent-free microwave extraction (V-SFME). *Chem. Eng. Commun.* **2020**. [CrossRef]

42. Bucić-Kojić, A.; Planinić, M.; Tomas, S.; Bilić, M.; Velić, D. Study of solid-liquid extraction kinectics of total polyphenols from grape seeds. *J. Food Eng.* **2007**, *81*, 236–242. [CrossRef]

43. Tomas-Barberán, F.A.; Cienfuegos-Jovellanos, E.; Marín, A.; Muguerza, B.; Gil-Izquierdo, A.; Cerdá, B.; Zafrilla, P.; Morillas, J.; Mulero, J.; Ibarra, A.; et al. A new process to develop a cocoa powder with higher flavonoid monomer content and enhanced bioavailability in healthy humans. *J. Agric. Food Chem.* **2007**, *55*, 3926–3935. [CrossRef] [PubMed]

44. Maisuthisakul, P. Antioxidant Potential and Phenolic Constituents of Mango Seed. *Kasetsart J. Nat. Sci.* **2009**, *297*, 290–297.

45. Crozier, S.J.; Hurst, W.J. Cocoa Polyphenols and Cardiovascular Health. In *Polyphenols in Human Health and Disease*; Watson, R.R., Preedy, V.R., Zibadi, S., Eds.; Elsevier Inc.: Amsterdam, The Netherlands, 2013; Volume 2, pp. 1077–1085. ISBN 9780123984562.

46. Racine, K.C.; Wiersema, B.D.; Griffin, L.E.; Essenmacher, L.A.; Lee, A.H.; Hopfer, H.; Lambert, J.D.; Stewart, A.C.; Neilson, A.P. Flavanol polymerization is a superior predictor of α-glucosidase inhibitory activity compared to flavanol or total polyphenol concentrations in cocoas prepared by variations in controlled fermentation and roasting of the same raw cocoa beans. *Antioxidants* **2019**, *8*, 635. [CrossRef] [PubMed]

47. Hasing, M.H. Estudio de la Variación en los Contenidos de Polifenoles y Alcaloides, en Almendras de Cacao por Efecto de los Procesos de Fermentación y Tostado. Ph.D. Thesis, Escuela Superior Politécnica de Chimborazo, Facultad de Ciencias, Escuela de Bioquímica y Farmacia, Riobamba, Ecuador, 2004.

48. Fernández-Romero, E.; Chavez-Quintana, S.G.; Siche, R.; Castro-Alayo, E.M.; Cardenas-Toro, F.P. The kinetics of total phenolic content and monomeric Flavan-3-ols during the roasting process of Criollo Cocoa. *Antioxidants* **2020**, *9*, 7–10. [CrossRef]

49. Niemenak, N.; Rohsius, C.; Elwers, S.; Ndoumou, D.; Liebereri, R. Comparative study of different cocoa (*Theobroma cacao* L.) clones in terms of their phenolics and anthocyanins contents. *J. Food Compos. Anal.* **2006**, *19*, 612–619. [CrossRef]

50. Rockenbach, I.I.; Jungfer, E.; Ritter, C.; Santiago-Schübel, B.; Thiele, B.; Fett, R.; Galensa, R. Characterization of flavan-3-ols in seeds of grape pomace by CE, HPLC-DAD-MS n and LC-ESI-FTICR-MS. *Food Res. Int.* **2012**, *48*, 848–855. [CrossRef]

51. Hurst, W.J.; Payne, M.J.; Miller, K.B.; Stuart, D.A. Stability of cocoa antioxidants and flavan-3-ols over time. *J. Agric. Food Chem.* **2009**, *57*, 9547–9550. [CrossRef] [PubMed]

52. Summa, C.; Raposo, F.C.; McCourt, J.; Scalzo, R.L.; Wagner, K.H.; Elmadfa, I.; Anklam, E. Effect of roasting on the radical scavenging activity of cocoa beans. *Eur. Food Res. Technol.* **2006**, *222*, 368–375. [CrossRef]

 *antioxidants*

Article

In Vitro Antioxidant Activity and FTIR Characterization of High-Molecular Weight Melanoidin Fractions from Different Types of Cocoa Beans

Joanna Oracz * and Dorota Zyzelewicz

Institute of Food Technology and Analysis, Faculty of Biotechnology and Food Sciences,
Lodz University of Technology, 4/10 Stefanowskiego Street, 90-924 Lodz, Poland; dorota.zyzelewicz@p.lodz.pl
* Correspondence: joanna.oracz@p.lodz.pl; Tel.: +48-42-631-3462

Received: 23 October 2019; Accepted: 14 November 2019; Published: 15 November 2019

Abstract: Melanoidins from real foods and model systems have received considerable interest due to potential health benefits. However, due to the complexity of these compounds, to date, the exact structure of melanoidins and mechanism involved in their biological activity has not been fully elucidated. Thus, the aim of this study was to investigate the total phenolic content, antioxidant properties, and structural characteristics of high-molecular weight (HMW) melanoidin fractions isolated by dialysis (>12.4 kDa) from raw and roasted cocoa beans of Criollo, Forastero, and Trinitario beans cultivated in various area. In vitro antioxidant properties of all studied HMW cocoa fractions were evaluated by four different assays, namely free radical scavenging activity against DPPH· and ABTS·$^+$ radicals, ferric reducing antioxidant power (FRAP), and metal-chelating ability. Additionally, the structure–activity relationship of isolated HMW melanoidin fractions were analyzed using attenuated total reflectance Fourier transform infrared spectroscopy (ATR-FTIR). The results show that roasting at a temperature of 150 °C and a relative air humidity of 0.3% effectively enhances the total phenolics content and the antioxidant potential of almost all HMW cocoa melanoidin fractions. The ATR-FTIR analysis revealed that the various mechanisms of action of HMW melanoidins isolates of different types of cocoa beans related to their structural diversity. Consequently, the results clearly demonstrated that HMW cocoa fractions isolated from cocoa beans (especially those of Criollo variety) roasted at higher temperatures with the lower relative humidity of air possess high antioxidant properties in vitro.

Keywords: melanoidins; *theobroma cacao* L.; total phenolic compounds; antioxidant capacity; metal-chelating ability; fourier transform infrared spectroscopy

---

1. Introduction

In recent years, a number of biological properties and potential health benefits of consuming cocoa-derived products have been investigated, e.g., antioxidant, anti-inflammatory, anti-carcinogenic, and antifungal properties. Much research has focused on the cocoa phenolic compounds, such as flavonoids (procyanidins, anthocyanins, flavonols, etc.), as a potential health-promoting compounds due to its antioxidant capacity and abundance in the cocoa beans. However, the bioactive compounds composition of cocoa beans is influenced by many factors including variety, climatic and agronomic conditions, post-harvest practices, and storage conditions [1–3]. Moreover, thermal processing may also cause a change in the level of phytochemicals in cocoa beans and their antioxidant capacity. Many studies have shown that phenolic compounds, which have been suspected as primarily factors responsible for the antioxidant properties of cocoa beans, are susceptible to degradation and oxidative condensation during thermal processing [2,4–7]. A recent study showed that, beside loss of phenolic compounds, roasting of cocoa beans induced only negligible changes in the total antioxidant capacity, probably due to the higher extractability of the cellular matrix compounds and/or formation of new antioxidants trough Maillard reactions, such as reductones and melanoidins [5,6,8]. Melanoidins are colored, nitrogen-containing, and polymeric compounds that form as a result of the final stage of the Maillard reactions. These compounds occur widely in many treated processed foods, such as coffee, bakery products, cooked potatoes, cocoa, roasted barley, and beef [9]. One of the most important properties that melanoidins contribute to foodstuffs is brown color. In addition, these macromolecules have considerable structural variability that result in diverse biological effects. The interest in the physiological role of melanoidins present in many heat-treated foods has increased dramatically over the last decade, particularly in relation to human health [2,9,10]. The biological effects exerted by melanoidins on the human body are thought to be strongly related to their ability the chelate metal cations, the capacity of scavenging superoxide anions and hydroxyl radicals, and the decomposing ability of hydrogen peroxide [10,11], which may be responsible for the antioxidant, anticancer, and antimicrobial properties [12]. Food melanoidins can also act as dietary fiber in the gastrointestinal tract and promote the growth of beneficial *Bifidobacteria* in the gut [9,11]. Current studies have shown that these biological functions are thought to be associated, at least in part, with the presence of phenolic compounds in the melanoidin structure [11,13]. Phenolic compounds, especially phenolic acids, are considered to contribute more to the healthful effects than the other constituents of melanoidins [13,14]. However, the exact mechanism of melanoidins antioxidant activity has not fully been elucidated to date. Moreover, despite numerous studies evaluating the biological and molecular properties of melanoidins obtained from both model systems and real foodstuffs, such as coffee, bread, honey, heated potato fiber, and malt [9,11,15], only very few authors have attempted to determine the biological activities of melanoidins isolated from cocoa beans and cocoa-derived products. Summa et al. [16,17] investigated the antibacterial, mutagenic, and radical-scavenging effects of four molecular weight fractions (>30, 30–10, 10–5, and <5 kDa) isolated by ultrafiltration from raw, pre-roasted (80–90 °C for 10 min), and roasted (130–160 °C for 15–20 min) cocoa beans. In those experiments, all high-molecular weight (HMW) fractions showed activity against the pathogenic bacteria *Enterobacter* and *Escherichia*. More recent studies have reported that LMW (<10 kDa), intermediate molecular weight (10–30 kDa), and HMW (>30 kDa) fractions isolated from cocoa powder by ultrafiltration have a dose-dependent inhibitory activity against α-glucosidase [18]. This activity was attributed to the presence of compounds in the low, intermediate, and high molecular weight fractions, including brown melanoidins, proteins, phenolic compounds, as well as polysaccharides, whether or not they were bound to the melanoidin skeleton or to unknown MRPs [18]. In our previous study, we investigated the effect of different roasting conditions, including temperature (110, 120, 135, or 150 °C) and relative air humidity (0.3% or 5.0%) on the physicochemical properties and the profiles of free phenolics and bound phenolics of high molecular weight Maillard reaction products isolated from Criollo, Forastero, and Trinitario beans from different regions of Africa [19,20]. In our previous work, we also demonstrated that the of the cocoa melanoidins are good sources of bound phenolics [19]. In that respect, it is noteworthy that the relation between the structure of these compounds and their health-related properties should be comprehensively investigated. Especially when, for some groups of people, cocoa beans represent a significant source of these compounds in their diet, the accurate assessment of the contribution of the chemical and structural properties of melanoidins to their antioxidant activity should be carried out.

In continuation with our previous study, this work was designed to carry out a comparative investigation of total phenolic content, antioxidant activity, and structural characteristic of HMW melanoidin fractions isolated from cocoa beans, both raw and roasted at different temperatures (110, 120, 135, and 150 °C) and relative air humidity levels (RH 0.3% and 5.0%) of three *Theobroma cacao* L. types. To the best of our knowledge, this is the first report concerning the determination of in vitro antioxidant activity and ATR-FTIR structural characterization of isolated HMW melanoidin fractions of Criollo, Forastero, and Trinitario beans from different regions of Africa.

2. Materials and Methods

2.1. Materials and Chemicals

Gallic acid, 6-hydroxy-2,5,7,8-tetramethylchroman-2-carboxylic acid (Trolox), 2,2′-azino-bis (3-ethylbenzothiazoline-6-sulfonic acid) diammonium salt (ABTS), 2,2-diphenyl-1-(2,4,6-trinitrophenyl) hydrazyl (DPPH), 2,4,6-tri(2-pyridyl)-s-triazine (TPTZ), sodium acetate, ferric chloride hexahydrate, ferrozine, and ammonium acetate were purchased from Sigma-Aldrich (St. Louis, MO, USA). All other reagents were of analytical grade and were purchased from Chempur (Piekary Slaskie, Poland). Ultrapure water (resistivity 18.2 MΩ cm), obtained from a Milli-Q purification system (Millipore, Bedford, MA, USA), was used for all analyses.

2.2. Plant Material

Analyses and experiments were conducted on samples of fermented and dried cocoa beans of the three main *T. cacao* groups: Criollo and Trinitario originating from Madagascar, and Forastero from Ghana. All cocoa beans were harvested at their technological maturity in 2013 and purchased from commercial sources. Cocoa fruit reaches physiological maturity around 5 months after flowering and is harvested at 5–6 months. Cocoa pods harvested at 5–6 months are considered to be technologically matured as the beans have developed optimum quality [21]. The description of cocoa beans is presented in Table 1.

Table 1. Selected physical characteristics of different cocoa varieties used in the study.

Variety	Country of Origin	Mean Size (Length × Width) (mm)	Mean Bean Weight (g)
Criollo	Madagascar	25.1 × 10.2	1.75 ± 0.07 [c]
Forastero	Ghana	22.3 × 8.1	1.32 ± 0.12 [a]
Trinitario	Madagascar	24.8 × 9.2	1.54 ± 0.08 [b]

Means sharing the different letters ([a–c]) are significantly different according to Tukey's HSD test at $p < 0.05$.

Raw cocoa beans of each group after removal of impurities and broken or chipped beans were convectively roasted in batches of 200 g in a tunnel with the forced air flow without circulation (adapted for either dry or humid air). Process air humidity was gradually increased using 4.0 MPa saturated steam from a generator. Roasting was performed at four temperatures (110, 120, 135, or 150 °C) and two relative air humidity levels (0.3% or 5.0%). The heat treatment parameters were chosen to obtain a range of roasted beans with acceptable physico-chemical and sensory properties. Usually, during roasting, the raw cocoa beans are exposed to temperatures that range from 135 to 150 °C, whereas the "fine or flavor" varieties require lower temperatures than the "bulk" ones [19]. It was terminated when bean moisture dropped to 2% as determined by drying until constant weight at 103 ± 2 °C. The time of thermal treatment was determined experimentally for each batch of cocoa beans, based on their initial water content and size. Roasting times at 150, 135, 120, and 110 °C were approximately 20, 40, 75, and 85 min, respectively. The application of higher relative air humidity prolonged the duration of thermal treatment. At the end of roasting, the beans were immediately cooled to approximately 20 °C for about 10 min. The roasted cocoa beans were kept in hermetically sealed plastic containers (500 g) and stored at −20 °C for subsequent analyses. The samples were analyzed within 6 to 12 h of storage. All roasting experiments were performed in duplicate for each cocoa type used.

2.3. Extraction and Isolation of Cocoa Melanoidin

The HMW melanoidin fraction was obtained from raw and roasted cocoa beans by dialysis according to our previously research [19]. Briefly, the cocoa beans were deshelled, ground, and defatted, and then extracted twice with 100 mL of water at 90 °C for 20 min in an orbital shaker (100× *g*). Subsequently, pooled extracts were cooled to room temperature and filtered through Whatman no.

4 filter paper to remove insolubles. An aliquot of the filtrate was dialyzed using a dialysis tubes (MW cut-off > 12.4 kDa, Sigma-Aldrich, Saint Louis, MO, USA) for 1 day under running tap water and for 2 days against 2000 mL of water at 4 °C with constant stirring. The water was changed 4–5 times until no further color was visible in the dialysate. The dialysate containing LMW compounds was removed. After dialysis, the retentate containing the HMW fraction was frozen at −20 °C and lyophilized (−50 °C, 0.9 MPa) using a DELTA 1-24LSC Christ freeze drier (Martin Christ, Osterode am Harz, Germany). All lyophilized HMW materials were then stored in plastic bags at −20 °C to prevent hydration until used.

2.4. Determination of Total Phenolic Content

Total phenolic contents (TPC) of HMW isolates from raw and roasted cocoa beans were determined using the Folin–Ciocalteu method, as described by Belščak et al. [22]. Briefly, 100 μL of the suitably diluted lyophilized HMW cocoa sample with high-purity deionized water (1.5 mg/mL) or blank was mixed with 4 mL of high-purity deionized water and 0.5 mL of the Folin–Ciocalteu reagent. After 3 min, 1 mL of 20% (w/v) Na_2CO_3 solution was added to the mixture. The final volume was adjusted to 10 mL with high-purity deionized water. The solution was then mixed vigorously and allowed to stand at room temperature in the dark place for 60 min. The absorbance of reaction mixture was measured at 765 nm using a UV-1800 spectrophotometer (Shimadzu, Tokyo, Japan). For each sample, experiments were conducted in triplicate. The results were expressed as mg gallic acid equivalents (GAE) per gram of lyophilized HMW cocoa fraction dry weight (mg GAE/g dw).

2.5. Determination of the Free Radical-Scavenging Capacity

The free radical-scavenging activity of the HMW isolates from raw and roasted cocoa beans was determined by the DPPH and ABTS assays as previously described by Oracz and Nebesny [8]. All analyses were carried out in triplicate, and the results obtained from the two tests were expressed as μmol Trolox equivalents per gram of lyophilized HMW cocoa fraction dry weight (μmol TE/g DW).

2.6. Determination of Ferric Reducing Antioxidant Power (FRAP)

The FRAP assay was performed according to the protocol described by Pastoriza and Rufián-Henares [23] with slight modification. The fresh working FRAP solution was prepared by mixing 25 mL of 300 mmol/L acetate buffer (pH 3.6), 2.5 mL of 10 mmol/L TPTZ in 40 mmol/L HCl, and 2.5 mL of 20 mmol/L $FeCl_3·6H_2O$ and then warmed at 37 °C for 30 min before using. Aliquots (0.1 mL) of the suitably diluted lyophilized HMW cocoa samples with high-purity deionized water (1.5 mg/mL) were mixed with 4 mL of fresh FRAP solution. The solution was then mixed vigorously and allowed to stand for 30 min at 37 °C. The absorbance of reaction mixture was measured at 593 nm using a UV-1800 spectrophotometer (Shimadzu, Tokyo, Japan). For each sample, experiments were conducted in triplicate. The results were expressed as μmol Trolox equivalents per gram of lyophilized HMW cocoa fraction dry weight (μmol TE/g dw).

2.7. Determination of Chelating Activity on Fe^{2+}

Chelating activity of the HMW isolates from raw and roasted cocoa beans was evaluated using the method of Gu et al. [24]. Briefly, 1 mL of the suitably diluted lyophilized HMW cocoa materials in high-purity deionized water (2 mg/mL) were added to 1.85 mL of high-purity deionized water and 50 μL of 2.0 mM $FeCl_2$. After mixing, the solution was allowed to stand at room temperature for 30 s, followed by the addition of 100 μL 5 mM ferrozine. The reaction mixture was then vortexed and left to stand at room temperature for 15 min. The absorbance of the solution was measured spectrophotometrically at 562 nm with UV-1800 UV-VIS Spectrophotometer (Shimadzu, Tokyo, Japan). Worth noting, that low absorbance of the resulting solution indicated a strong ferrous ion chelating ability. A reaction mixture containing 1 mL of high-purity deionized water instead of sample solution served as the blank. The results are expressed as the percentage chelating activity (%).

2.8. Attenuated Total Reflection Fourier Transform Infrared Spectroscopy

The infrared spectroscopy is based on the absorption of radiation due to vibrations bonds of molecules [25,26]. The FTIR spectra of lyophilized HMW cocoa melanoidin fractions were obtained using an infrared Fourier transform spectrometer, model IRTracer-100 (Shimadzu, Tokyo, Japan) equipped with an attenuated total reflection (GladiATR) accessory with diamond crystal (PIKE Technologies, Inc., Madison, Wisconsin, USA) at room temperature. The spectral range was 400–4000 cm^{-1} with 40 scans and a resolution of 4 cm^{-1}. Around 5 mg of lyophilized HMW cocoa fractions was deposited on the diamond platform prior to measurement. Background and sample spectra were acquired at 4 cm^{-1} resolution with 40 scans from 400 to 4000 cm^{-1}

2.9. Statistical Analysis

The results are presented as mean ± standard deviations of three replicates. Statistical tests were evaluated by using the Statistica 13.0 software (StatSoft, Inc., Tulsa, OK, USA). The obtained data were tested for normal distribution (Shapiro-Wilk test) and equal variances (Levene's test). As all values showed normal distribution and homogeneity of variance, considering $p \geq 0.05$, the data were subjected to the analysis of variance (ANOVA). The effects of variety, roasting temperature, roasting air humidity, and their interaction on total phenolic content and antioxidant activity of HMW melanoidin fractions were tested by means of two-way ANOVA. The significant differences among the means were estimated through Tukey's HSD test. For all statistical analysis, $p < 0.05$ was considered as statistical significance. The error bars in all figures correspond to the standard deviations. The correlation coefficients between investigated parameters were assessed by means of the Pearson correlation test using Microsoft Office Excel 2016 (Microsoft Corporation, Redmond, WA, USA).

3. Results and Discussion

The present study is a continuation of our previous work [19,20] exploring the physicochemical properties and the profiles of free and bound phenolics of HMW melanoidin fractions of different *T. cacao* groups and origins. In this work, the total phenolic content and the antioxidant activity of the HMW melanoidin fractions (>12 kDa) from raw and roasted, at different temperatures and relative air humidities, cocoa beans of different groups were evaluated by using different in vitro spectrophotometric assays.

3.1. Total Phenolics Content

The total phenolics content (TPC) of HMW melanoidin fractions from the three cocoa types is shown in Figure 1. The data were expressed as milligrams of gallic acid equivalents (GAE) per gram dry weight (mg GAE/g dw). In this study, significant differences ($p < 0.05$) in TPC of the HMW fractions were found between the three cocoa types. The HMW fractions from both raw and roasted Criollo beans were characterized by the highest TPC, ranging from 121.99 to 149.90 mg GAE/g DW. Less phenolics were observed in HMW isolates from Forastero beans (117.46–139.12 mg GAE/g DW), while HMW materials from Trinitario beans contained the lowest amounts of these compounds (93.02–129.00 mg GAE/g DW). In order to assess the influence of cocoa variety, roasting temperature, roasting air humidity, and their interactions on TPC content of HMW melanoidin fractions, a two-way ANOVA was carried out. The results showed that the cocoa variety and roasting process parameters significantly affected the TPC of HMW melanoidins isolated from the studied cocoa beans ($p < 0.01$). Generally, more phenolics were contained in the almost all of HMW fractions isolated from roasted beans than from the raw ones. The greatest increase in TPC (by 25.3–38.1% of initial value) of the HMW fractions was observed when cocoa beans of the Trinitario type were roasted at 110 °C. Moreover, it was also noticed that thermal processing of Criollo and Forastero beans at 150 °C with RH of 0.3% led to the highest increase in the TPC (by 18.4–22.9% of initial value). As can be seen from the results, the changes in the TPC of the HMW fractions also depended on the RH and in almost all samples were slightly less advanced when the air humidity was increased from 0.3% to 5.0%. Thus, it was confirmed that influence of the roasting parameters on TPC of the HMW fractions is complex, and huge differences are observed depending on applied cocoa bean variety and process conditions, like temperature and relative air humidity level. According to Pérez-Martínez et al. [27], the Folin–Ciocalteau assay measures the reducing capacity of a sample, and the results of TPC may be affected by the presence of other electron donors, such as non-phenolic substances and nitrogen-containing compounds. Thus, the increase in the level of compounds able to react with the Folin–Ciocalteu reagent could be linked to the formation of new substances, especially reductones, during the roasting of cocoa beans. These compounds, which are formed in the advanced and final stages of Maillard reaction, can act as reducing agents, due to the presence of hydroxyl and pyrrole groups [24,28].

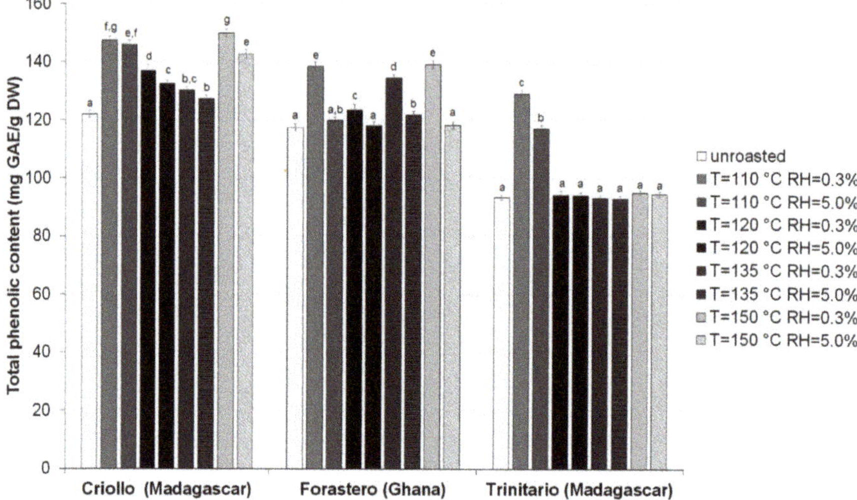

Figure 1. Total phenolics content of melanoidin fractions isolated from raw and roasted, at different at different temperatures and relative air humidities, cocoa beans of different groups. Results are presented as means ± SD from triplicate assays. Bars with the same lowercase letter (a–g) within each variety do not differ significantly according to Tukey's HSD test at $p < 0.05$.

Summa et al. [16] reported that pre-roasting of cocoa beans (at 80–90 °C for 10 min) significantly reduced the concentration of Folin–Ciocalteu reactive substances in the 30–10 kDa fraction while accurate roasting (at 130–160 °C for 15–20 min) increased their content. As suggested by these authors, this phenomenon might be caused by formation of reducing substances in the 30–10 kDa fraction during roasting at high temperatures. They also found that the level of reducing compounds in the >30 kDa fraction increased significantly during cocoa beans pre-roasting but declined markedly after accurate roasting. In addition, the observed behavior can be a consequence of the incorporation of phenolic compounds into the structure of melanoidins during heat treatment [15,29]. The oxidized polyphenols may react with thiol groups on amino acids, peptides, and proteins or nucleophilic amines via Michael-type addition or Schiff base reaction to form thiol–quinone and amine–quinone adducts or benzoquinone imines [15,30–32]. Our previous studies have shown that roasting of cocoa beans of different types at temperatures ranging from 110 to 150 °C led to a marked increase in the bound phenolics content as compared to HMW fractions from unroasted beans [20]. Therefore, we can concluded that significant differences in the TPC observed between the HMW cocoa melanoidin fractions appeared to be brought about by the structural changes of their active components, involving oxidation of phenolic compounds, their condensation with proteins and polysaccharides, as well as cross-linking and polymerization of low molecular weight MRPs during roasting, which decide their electron donating abilities [20,31,33].

3.2. Free Radicals-Scavenging Capacity

To make a comprehensive evaluation on the antioxidant effect of samples, it is necessary to employ different methods due to the fact that this assays exhibit different mechanisms of by which its antioxidant activity takes place. In vitro free radical scavenging activities of the HMW melanoidin fractions from raw and roasted at various conditions cocoa beans of different *T. cacao* groups and origins were assessed against DPPH and ABTS radicals (Table 2).

Table 2. The free radical scavenging capacity (2,2-diphenyl-1-(2,4,6-trinitrophenyl) hydrazyl (DPPH) and 2,2′-azino-bis (3-ethylbenzothiazoline-6-sulfonic acid) diammonium salt (ABTS)) and ferric reducing ability (FRAP) of high-molecular weight (HMW) fractions isolated from raw and roasted at different conditions cocoa beans of Criollo, Forastero, and Trinitario groups originating from various geographical regions.

Varieties (Country of Origin)	Roasting Conditions	DPPH	ABTS	FRAP
Criollo (Madagascar)	unroasted	726.26 ± 1.53 [a]	849.84 ± 1.65 [c]	622.42 ± 1.71 [a]
	T = 110 °C, RH = 0.3%	902.61 ± 1.65 [c]	920.74 ± 1.68 [d,e]	846.75 ± 1.67 [b,c]
	T = 110 °C, RH = 5.0%	826.77 ± 1.74 [b]	943.17 ± 1.72 [e]	869.94 ± 1.59 [c]
	T = 120 °C, RH = 0.3%	1172.37 ± 1.76 [e]	1127.53 ± 1.54 [g]	915.42 ± 1.62 [d]
	T = 120 °C, RH = 5.0%	1074.21 ± 1.71 [d]	1086.72 ± 1.65 [f]	841.03 ± 1.59 [b]
	T = 135 °C, RH = 0.3%	1384.97 ± 1.68 [g]	805.94 ± 1.59 [b]	1077.18 ± 1.72 [g]
	T = 135 °C, RH = 5.0%	1250.47 ± 1.72 [f]	772.74 ± 1.73 [a]	1036.51 ± 1.63 [f]
	T = 150 °C, RH = 0.3%	1475.61 ± 1.75 [h]	903.34 ± 1.68 [d]	987.45 ± 1.55 [e]
	T = 150 °C, RH = 5.0%	1411.21 ± 1.62 [g]	897.71 ± 1.64 [d]	904.74 ± 1.49 [d]
Forastero (Ghana)	unroasted	689.16 ± 1.35 [a]	806.22 ± 1.67 [a]	553.90 ± 1.49 [a]
	T = 110 °C, RH = 0.3%	1046.65 ± 1.86 [d]	917.64 ± 1.59 [d,e]	785.85 ± 1.52 [d]
	T = 110 °C, RH = 5.0%	938.30 ± 1.88 [b]	857.05 ± 1.78 [b]	714.41 ± 1.68 [b]
	T = 120 °C, RH = 0.3%	1124.59 ± 1.79 [e]	937.50 ± 1.72 [e]	825.15 ± 1.71 [e]
	T = 120 °C, RH = 5.0%	989.95 ± 1.85 [c]	903.18 ± 1.65 [d]	750.13 ± 1.63 [c]
	T = 135 °C, RH = 0.3%	1224.05 ± 1.89 [g]	819.04 ± 1.68 [a]	947.46 ± 1.70 [h]
	T = 135 °C, RH = 5.0%	1172.50 ± 1.67 [f]	826.16 ± 1.57 [a]	911.02 ± 1.56 [g]
	T = 150 °C, RH = 0.3%	1309.52 ± 1.91 [h]	897.62 ± 1.64 [c,d]	817.99 ± 1.59 [e]
	T = 150 °C, RH = 5.0%	1177.01 ± 1.84 [f]	877.13 ± 1.71 [b,c]	850.71 ± 1.58 [f]
Trinitario (Madagascar)	unroasted	640.43 ± 1.29 [a]	680.90 ± 1.56 [a]	440.49 ± 1.42 [a]
	T = 110 °C, RH = 0.3%	1024.17 ± 1.45 [d]	833.81 ± 1.71 [c]	728.69 ± 1.71 [g]
	T = 110 °C, RH = 5.0%	1006.96 ± 1.37 [d]	874.36 ± 1.39 [d]	698.13 ± 1.64 [e]
	T = 120 °C, RH = 0.3%	1209.25 ± 1.39 [f]	976.25 ± 1.48 [g]	624.41 ± 1.72 [c]
	T = 120 °C, RH = 5.0%	1107.30 ± 1.27 [e]	948.24 ± 1.53 [f]	590.76 ± 1.59 [b]
	T = 135 °C, RH = 0.3%	938.00 ± 1.68 [c]	897.84 ± 1.67 [e]	779.49 ± 1.76 [i]
	T = 135 °C, RH = 5.0%	834.00 ± 1.46 [b]	894.13 ± 1.56 [d,e]	768.42 ± 1.36 [h]
	T = 150 °C, RH = 0.3%	1189.09 ± 1.24 [f]	714.99 ± 1.41 [b]	654.89 ± 1.52 [d]
	T = 150 °C, RH = 5.0%	1085.65 ± 1.37 [e]	689.24 ± 1.72 [a]	584.27 ± 1.65 [b]
Two-way ANOVA analysis			Significance	
	Variety [V]	**	*	***
	Roasting temperature [RT]	***	***	***
	Roasting air humidity [RH]	**	ns	**
	Interactions of V × RT × RH	***	***	***

T, temperature. RH, relative air humidities. Data are presented as mean ± SD of three replications. The means followed by the same lowercase letter ([a–h]) within each variety in the same column do not differ significantly according to Tukey's HSD test at $p < 0.05$. Significance: * $p < 0.05$; ** $p < 0.01$; *** $p < 0.001$; ns: not significant. DPPH, the DPPH radical-scavenging capacity expressed in μmol Trolox equivalents per gram of HMW fraction dry weight (μmol TE/g DW); ABTS, the ABTS radical-scavenging activity expressed in in μmol Trolox equivalents per gram of HMW fraction dry weight (μmol TE/g DW). FRAP, the ferric reducing antioxidant power expressed in μmol Trolox equivalents per gram of HMW fraction dry weight (μmol TE/g DW).

The ABTS$\cdot^+$ and DPPH assays are widely used methods for the assessment of the antioxidant activities of many vegetable or food matrices [34–37]. These methods are both based on quenching of stable colored radicals and show the free radical quenching activity of antioxidants even when present in complex biological matrices such as plant or food preparations (extracts or fractions). The values of antioxidant potential obtained in both assays were equal or higher than those reported by other authors for the HMW melanoidin fractions of cocoa beans, chocolate, or even coffee [16,23,38], meaning that all of studied HMW cocoa melanoidin samples exhibited good DPPH and ABTS radical scavenging activities.

However, there were significant differences between HMW melanoidin fractions from both the raw and roasted cocoa beans regarding their antioxidant abilities. Two-way ANOVA revealed significant effect of cocoa variety ($p < 0.01$), roasting temperature ($p < 0.001$), roasting air humidity ($p < 0.01$), and interaction of variety and roasting conditions ($p < 0.001$) for the DPPH radical-scavenging activity (Table 2). In addition, ABTS scavenging capacity were significantly affected by variety, roasting temperature, and interaction of variety and roasting conditions, but the levels of relative humidity did not have significant influence. The HMW materials from unroasted Criollo beans demonstrated the highest scavenging capacity against both DPPH and ABTS radicals (726.26 and 849.84 µmol TE/g dw, respectively). Compared with HMW melanoidins from Criollo beans, the HMW fraction from unroasted Forastero beans showed the significantly lower ($p < 0.05$) activity against DPPH (689.16 µmol TE/g dw) but the similar against ABTS radical cations (806.22 µmol TE/g dw). The lowest radical scavenging abilities against the ABTS and DPPH assays was exhibited by HMW isolates from unroasted Trinitario beans, which is consistent with the results of TPCs discussed above. These results showed that, depending on the cocoa type and roasting conditions, an increase or decrease in the antioxidant capacity of HMW melanoidin fractions was observed. Roasting of all types of cocoa beans at temperatures ranging from 110 to 150 °C caused a significant increase ($p < 0.05$) in the DPPH radical-scavenging activity of HMW fractions. However, the changes in the free radical-scavenging activity of HMW fractions were considerably smaller at the higher RH. Interestingly, we observed that HMW fractions from Criollo beans roasted at 150 °C with lower RH exhibited a higher antioxidant activity by DPPH scavenging assay (1475.61 µmol TE/g dw) compared to all other melanoidin samples. Roasting of Forastero and Trinitario beans at temperatures ranging from 110 to 150 °C caused a significant increase ($p < 0.05$) in the ABTS radical-scavenging activity of their HMW melanoidin fractions. It was also observed that the HMW fractions from Criollo beans, roasted at temperatures of 110, 120, and 150 °C, displayed the significantly increased ABTS scavenging capacity, while after roasting at 135 °C, exhibited a slightly lower antioxidant activity by this method. Nevertheless, among HMW melanoidin fractions isolated from roasted beans of the three cocoa varieties, the highest ABTS radical-scavenging capacity (1172.37 µmol TE/g dw) was exhibited by the fraction from Criollo beans roasted at 120 °C and RH of 0.3%. Based on the results presented in Table 2, it is possible to observe that, in almost all HMW fractions, antioxidant activity is higher in the DPPH method (except HMW fractions of Criollo beans roasted at 110 °C). This behavior is similar to the observed by other researchers [23] who have observed a stronger scavenging activity against DPPH· than ABTS$\cdot^+$ radicals in the case of chocolate melanoidins. Our results corroborated also with those reported by Summa et al. [16], who showed that the >30 kDa fraction obtained from pre-roasted cocoa beans had higher DPPH and ABTS radical-scavenging activity compared to the >30 kDa fractions of raw and roasted beans. The differences between the ABTS and DPPH radical-scavenging activity could be explained by the wide variety of chemical mechanisms involved in the antioxidant activity of HMW cocoa melanoidin fractions, as it was mentioned earlier. The reaction mechanism with DPPH· radical involve transfer of a hydrogen atom, while the reactions with ABTS$\cdot^+$ radicals involve electron transfer process [39]. The observed differences indicate the complexity of the mechanism of action of melanoidins formed in real food. The increase in the antioxidant capacity of the HMW fractions can be a consequence of accumulation of high molecular weight MRPs-like melanoidins during roasting. In addition, the observed behavior can be explained by the presence of residues of certain active compounds, containing more than one active group (OH or NH_2), such as phenolic compounds, quinones, and low molecular weight MRPs, in the HMW materials. These compounds might be attached to the structure of melanoidins via non-covalent bonds and influence their biological properties [4,5,18]. Although the content of free phenolic compounds in HMW fractions of roasted cocoa beans decreased [19], it is possible that it could be explained by the presence of quinones generated by oxidation of these compounds that spontaneously form covalent bonds to functional groups of melanoidins during roasting [4,40]. We suppose that the radical-scavenging activity of tested HMW fractions from cocoa beans depends on the structure and the number of the included active group (OH or NH_2). This was a clear indication that the synergistic effect between different bioactive compounds present in the structure of melanoidins could determine their biological properties. Since, the HMW cocoa melanoidins have the ability to scavenge free radicals, thereby preventing lipid peroxidation chain reactions that cause damage a wide range of molecules found in living cells, they could serve as potential nutraceuticals and functional ingredients.

3.3. Ferric Reducing Antioxidant Power

The reducing capacity may serve as a significant indicator of potential antioxidant activity. Thus, the HMW melanoidin fractions of unroasted and roasted cocoa beans of tree different *T. cacao* beans was estimated for their ability to reduce TPTZ–Fe (III) complex to TPTZ–Fe (II). In order to identify the influence of cocoa variety, roasting temperature, roasting air humidity, and their interactions on ferric reducing antioxidant power (FRAP), a two-way ANOVA was performed. As demonstrated in Table 2, significant differences ($p < 0.05$) in the reducing properties of HMW fractions were found between the three cocoa types. The HMW fractions from Criollo beans exhibited the highest FRAP values, ranging from 622.42 to 1077.18 µmol TE/g dw, followed by isolates of Forastero beans (533.90–947.46 µmol TE/g dw), while HMW materials of Trinitario beans had the lowest reducing power (440.49–779.49 µmol TE/g dw). The results also showed that the roasting process parameters significantly ($p < 0.05$) affected the reducing properties of HMW materials isolated from the studied cocoa beans. Generally, all of HMW fractions isolated from unroasted beans showed a lesser reducing power compared to the isolates from the roasted samples. The greatest increase in the ferric reducing ability (by 71.1%–77.0% of initial value) of HMW fractions was observed when cocoa beans of all studied cocoa groups were roasted at 135 °C with RH of 0.3%. As can be seen from the results, the changes in the reducing power of HMW fractions depended on the RH and in almost all samples was slightly more pronounced when the air humidity was decreased from 5.0% to 0.3%. We can conclude that all studied HMW cocoa melanoidin fractions behaved as reductants and inactivators of oxidants due to the presence of electron-donors in their structure. The observed increase in the reducing capacity of HMW cocoa fractions could be linked to the formation of new substances able to donate electrons or to terminate radical chain reactions, during roasting of cocoa beans. As described above, the HMW cocoa melanoidin fractions contained different amounts of free and bound phenolic compounds and the presence of these polyphenols along with compounds formed in the advanced and final stages of Maillard reaction could contribute to the antioxidant properties observed by FRAP method. These residues due to the presence of hydroxyl and pyrrole groups can act as reducing agents via their redox potential of transferring electrons [24,33]. In a study performed by Summa et al. [16], it was observed that roasting of cocoa beans (at 130–160 °C for 15–20 min) significantly increased the concentration of reducing substances in the 30–10 kDa fraction. They also found that the level of reducing compounds in the >30 kDa fraction increased significantly during cocoa beans pre-roasting but declined markedly after roasting.

3.4. The Weighted Average Antioxidant Capacity

Our results suggest that the antioxidant capacity of HMW fractions from raw and roasted beans depend on the cocoa type and roasting conditions, wherein no significant correlation was observed between DPPH, ABTS, and FRAP values. This may be explained by the fact that different constituents in the HMW fractions would have different mechanisms of action regarding their antioxidant activities.

According Tabart et al. [41], the weighted average of the results obtained by the different assays should be calculated to get an overall impression of the antioxidant potential of the samples. Therefore, the results of antioxidant capacity of HMW fraction obtained by the specific assay (DPPH, ABTS, and FRAP) was divided by the average activity of the all HMW samples by the same assay, and then the calculated values in each assay were summarized and divided by the number of assays used (three in our case). The weighted average antioxidant capacity of each HMW samples are shown in Figure 2. The results of the present study show that cocoa beans roasting temperatures of 110–150 °C cause significant rise in the weighted average antioxidant capacity (WAAC) of the HMW materials. The highest WAAC values were obtained at a roasting temperature of 150 °C and RH of 0.3% (in the case of Criollo and Forastero groups) and a roasting temperature of 120 °C and RH of 0.3% (in the case of Trinitario group). It was also observed that the HMW fractions of both raw and roasted Criollo beans exhibited the highest WAAC values. The HMW fractions derived from almost all beans of Forastero type has a slightly higher WAAC compared to the samples of Trinitario type (except for beans roasted at 110 °C and RH of 5.0%). This is reflected in the results of the two-way ANOVA (Figure 2) indicating that WAAC was significantly affected by the cocoa variety ($p < 0.001$), roasting temperature ($p < 0.001$), roasting air humidity ($p < 0.01$), and their interactions ($p < 0.001$).

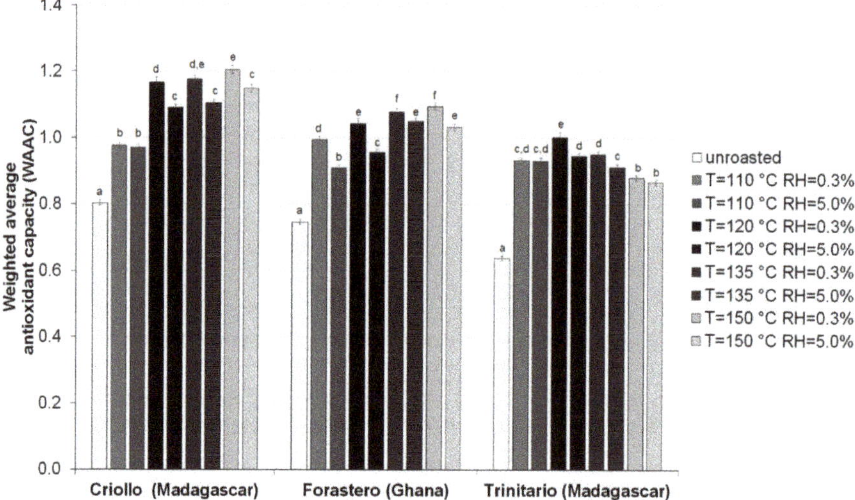

Figure 2. The weighted average antioxidant capacity of HMW fractions isolated from raw and roasted, at different at different temperatures and relative air humidities, cocoa beans of different groups. Results are presented as means ± SD from triplicate assays. Bars with the same lowercase letter (a–f) within each variety do not differ significantly according to Tukey's HSD test at $p < 0.05$.

3.5. Chelating Activity on Fe^{2+}

The changes in ferrous ion chelating activity of HMW fractions, caused by roasting of beans of all tested cocoa groups, are presented in Figure 3. All HMW isolates obtained from raw and roasted cocoa beans were able to chelate ferrous ion (Fe^{2+}), being the most powerful pro-oxidants among various species of metal ions. The ferrous ion in the Fenton reaction can catalyze the generation of potentially toxic reactive oxygen species (ROS), such as hydroxyl radicals ($\cdot$OH) that initiate lipid peroxidation [28]. It was also found that, similar to the WAAC, the HMW fractions from raw Criollo beans showed the higher chelating activity (55.16%) than those derived from raw Forastero and Trinitario beans (53.31% and 43.39%, respectively). To the best of our knowledge, this is the first study showing the interplay between roasting conditions and the iron chelating activity of HMW fractions isolated from different groups of cocoa beans. Two-way ANOVA revealed that the effect of cocoa variety, roasting conditions, and their interactions was highly significant ($p < 0.001$). The results of this study demonstrate that the chelating activity of HMW fractions isolated from beans of the three cocoa types was increased by roasting at temperatures in the range of 110 to 150 °C, which may be ascribed to structural changes in melanoidins and phenolic compounds. Moreover, the changes in the iron binding ability greatly depended on the roasting air humidity (0.3–5.0%) and were considerably smaller when the RH was elevated. The chelating activity of the HMW isolates obtained from roasted samples varied from 45.21% to 92.07%. The increase in the metal-chelating activity was more pronounced in HMW isolates from Forastero and Trinitario beans, while the lowest changes were observed in the Criollo samples. Irrespective of the cocoa type, the greatest increase in the iron chelating ability occurred when cocoa beans were roasted at 150 °C and the lower relative humidity of air (RH of 0.3%) whereas the lowest rise was caused by roasting at 110 °C with humid air (RH of 5.0%). These results are consistent with the described above differences in the free radical scavenging activities and the reducing properties of HMW cocoa. However, unlike to the WAAC, the melanoidin fraction of Forastero beans roasted at higher temperatures exhibited the highest ferrous ion chelating potential, compared to other cocoa groups (even in the relation to Criollo samples).

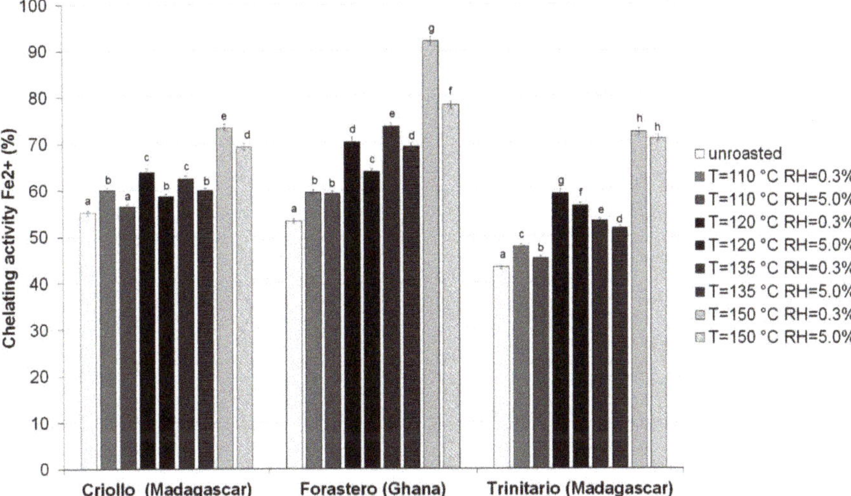

Figure 3. Total phenolics content of HMW fractions isolated from raw and roasted, at different at different temperatures and relative air humidities, cocoa beans of different groups. Results are presented as means ± SD from triplicate assays. Bars with the same lowercase letter (a–h) within each variety do not differ significantly according to Tukey's HSD test at $p < 0.05$.

Such differences could be explained by variations in the composition of studied HMW melanoidin fractions that are strongly determined by cocoa variety and roasting conditions [19,20]. In addition, this phenomenon suggests that Maillard reactions that took place during roasting of cocoa beans at the higher temperatures resulted in formation of high molecular weight brown melanoidins, exhibiting the high iron chelating activity. Our findings confirm recently reported observation that MRPs with metal-chelating ability are generated in model systems due to heat treatment. Gu et al. [24] found that high molecular weight MRPs had higher metal-chelating potential than low molecular weight MRPs. They also suggest that the metal-chelating activity of MRPs is possibly affected by the presence of the hydroxyl or pyrrole groups in their structures. This was a clear evidence that the studied HMW cocoa melanoidins may prevent against oxidative damage by sequestering Fe (II) ions that participate in transition metal ions-catalyzed hydrogen peroxide decomposition with generation of hydroxyl radicals.

3.6. Characterization of Cocoa Melanoidins by Fourier Transform Infrared Spectroscopy

To provide a comprehensive explanation about the antioxidant properties of HMW cocoa fractions, it was necessary to precisely characterize the chemical structure of these materials. We report here for the first time the ATR-FTIR spectroscopic characterization of HMW cocoa melanoidins. The FTIR spectra of HMW isolates obtained from raw and roasted cocoa beans were collected with FTIR spectrometer equipped with an ATR sample accessory. The ATR-FTIR spectra (within the 400 cm^{-1} to 4000 cm^{-1} wavenumber region) of the cocoa melanoidins are given in Figure 4.

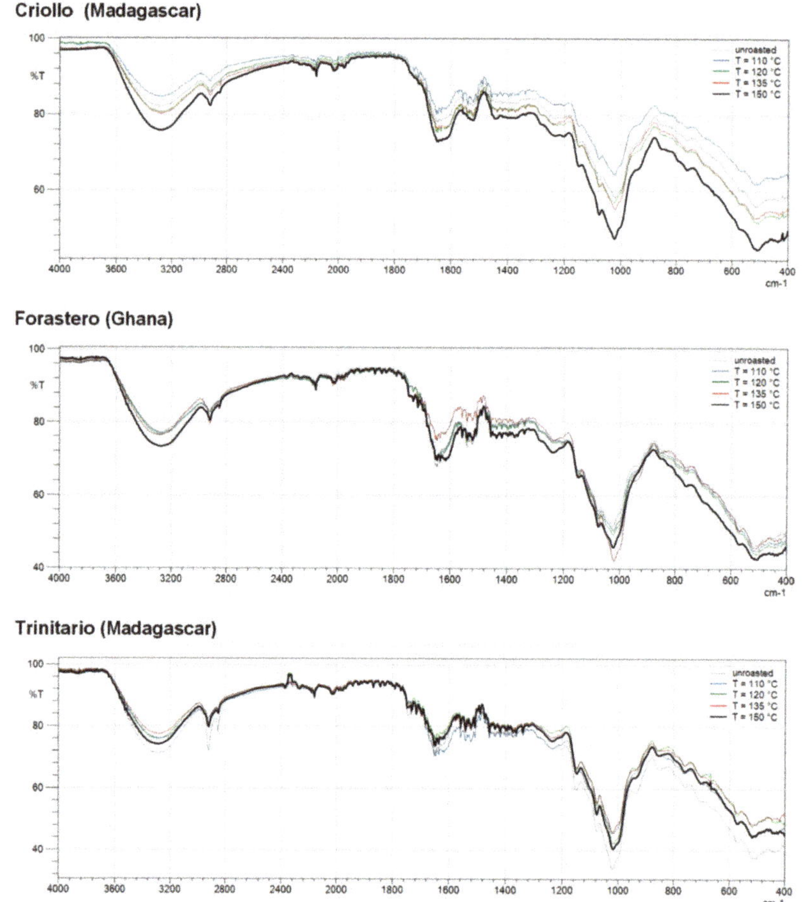

Figure 4. Fourier transform infrared spectroscopy (FTIR) spectra of HMW fractions isolated from raw and roasted, at different at different temperatures and relative air humidities, cocoa beans of different groups.

The heterogenic nature of the cocoa melanoidins has been confirmed by Fourier transform infrared spectra of HMW fractions isolated from different types of *T. cacao* beans, which exhibit prominent absorption bands in the broad region 400–4000 cm^{-1}, the characteristic for various classes of compounds. Common features as well as particular vibrations, specific to from phenolic compounds and its derivatives, polysaccharides, and proteins, are found in the spectra. The ATR-FTIR spectra of all HMW cocoa melanoidin fractions contains a wide band at 3350 cm^{-1} that belongs to the H-bond stretching vibrations of O–H hydroxyl groups, strong stretching band at about 1650 cm^{-1} assigned to the double bond stretching of carbonyl (C=O), C=C, or C=N [42] and characteristic absorption bands appeared between 900–600 cm^{-1} due to the stretching vibrations of the entire anhydroglucose ring. All mentioned bands confirms the presence of a phenol group sensitive to hydrogen bonding [25]. Moving forward, it confirms that more phenolic compounds bound to melanoidins might also contribute to the observed strong antioxidant capability of HMW cocoa fractions. The antioxidant activity of phenolic compounds is attributed to its molecular structures, particularly the number and positions of the hydroxyl groups, and the nature of substitutions on the aromatic rings [43]. Therefpre, more active antioxidant compound possesses more hydroxyl groups. The changes in the spectra gradually occur, depending on the cocoa type and roasting conditions. Our results showed that thermal treatment temperatures between 110 and 150 °C generally caused an increase in intense of the peak of HMW fractions isolated from cocoa beans (except for Trinitario samples). The peaks at 2922 and 2850 cm^{-1} in the FTIR spectra were attributed to the stretch of the C–H of aromatic ring and would be due to stretching vibrations of CH$_2$ and CH$_3$ groups that can originate from fatty acids present in the cocoa melanoidin fractions [25]. A igher peak in the region 2920 cm^{-1} clearly reflects adsorption of aliphatic compounds in the melanoidins structure. Some reports revelated that interactions between reactive carbonyl compounds arising from lipid oxidation reactions and amino acids or proteins might play an important role in the formation of brown HMW macromolecular compounds upon high-temperature processing of cocoa beans [44,45]. Bands in the range of 1616–1690 cm^{-1} were ascribed to N–H bending vibrations from amine or amide groups, and C=O stretching vibrations from flavonoids, phenolic acids and its derivatives, quinones, and lipids [45,46]. The band centered at about 1650 cm^{-1} mainly corresponds to the C=O stretching vibrational mode of the different structures of the protein backbone. The band at 1515 cm^{-1} was attributed to C=C stretching vibrations from aromatic rings of phenolic compounds [25,26,45]. Mot [46] noticed that the intensity of the peak at 1630 cm^{-1} is closely correlated with the antioxidant activity of plant extracts. In this study, in the FTIR spectrum of HMW fractions of Trinitario beans, the intensity of the 1630 cm^{-1} peak is significantly lower compared to the Criollo and Forastero melanoidins. This behavior corroborates the results obtained in DPPH· and ABTS scavenging assays and could be a consequence of the higher number of phenolic groups in the HMW melanoidin fractions of Criollo and Forastero groups compared to those of Trinitario type. Moreover, the relative intensity of this peak increases during roasting of almost all cocoa beans, which may be the result of adsorption of phenolic compounds and/or its derivatives in the melanoidin structures. This was a clear evidence that the studied HMW melanoidin fractions possessed residues (OH or NH$_2$ groups) that could act simultaneously as hydrogen and electron donors. Our previous study on UHPLC-DAD-ESI-HR-MS^n analysis of HMW melanoidin fractions derived from two different types of cocoa beans revealed that both free and bound phenolic compounds, including three flavan-3-ols, seven phenolic acids, one phenolic aldehyde, and four N-phenylpropenoyl-L-amino acids (NPAs), are present in these fraction [20]. Most of the phenolics in all the HMW melanoidin fractions were present in the bound form. It was also found that HMW fractions obtained from roasted cocoa beans had the higher content of bound phenolics than those from unroasted cocoa beans. This observation is consistent with the results of other authors who showed that phenolic compounds can be incorporated into the melanoidin skeleton during coffee roasting [15]. The spectral region from 1400 to 650 cm^{-1} is called the fingerprint region. This region of the infrared spectrum contains vibrations that are specific for the large number of infrared bands, including C–O, C–C, and C–N single bond stretches, C–H bending vibrations, and some bands due to benzene rings [25,26]. The peaks in the range of 1020–1060 cm^{-1} corresponded to C–O–C and C–O stretching vibrations of the glycoside linkage and C–O bond stretching vibration in glycerol. According to the literature, the occurrence of this band is an evidence of polysaccharide and lipids in samples [25,26].

Our results demonstrate that ATR-FTIR spectroscopy may be used as a direct and nondestructive method for the rapid investigation of the structural characteristics and functional properties of the HMW melanoidin fractions. According to our findings, the presence of different types of compounds in HMW melanoidin fractions is responsible for their bioactive properties (e.g., reducing power, antioxidant capacity, chelating activity) and its different mechanisms of action.

4. Conclusions

In conclusion, the present study, to the best of our knowledge, was the first time a comprehensive study was carried out on the total phenolic content and antioxidant activities, as well as structure–activity relationships of HMW cocoa melanoidin fractions isolated from raw and roasted, under different temperature and relative air humidity conditions, cocoa beans of different *T. cacao* groups. The results showed that the cocoa type and roasting conditions affect the total phenolic content and antioxidant properties of HMW melanoidin fraction isolated from the studied beans. The ATR-FTIR analysis revealed the presence of different bioactive compounds with various mechanism of action in HMW cocoa melanoidin fractions. We found that, both for TPC and in vitro antioxidant activity, HMW cocoa melanoidin fractions of Criollo beans showed significantly higher values than those of Fosrastero and Trinitario samples. Moreover, we observed that that roasting at higher temperatures with the lower relative humidity of air effectively enhances in vitro antioxidant potential of almost all HMW fractions isolated from cocoa beans. Consequently, irrespective of the cocoa type, the thermal processing at 150 °C and RH of 0.3% can be recommended to obtain HMW materials with the highest total phenolic content and strong in vitro antioxidant potential. This fact could indicate that the thermal processing of cocoa beans enhances the concentration of bioactive molecules in HMW fractions that contribute to the antioxidant response observed in in vitro tests. Our findings suggest, in general, that optimization of the roasting conditions and choosing an appropriate cocoa variety may provide functional advantages by enhancing in vitro antioxidant properties of HMW fractions of cocoa beans.

Author Contributions: Conceptualization, J.O.; formal analysis, J.O.; funding acquisition, J.O.; investigation, J.O. and D.Z.; methodology, J.O.; resources, J.O. and D.Z.; writing—original draft preparation, J.O.

Funding: This research was funded by NATIONAL SCIENCE CENTRE POLAND, grant number UMO-2012/05/N/NZ9/01399.

Conflicts of Interest: The authors declare no conflict of interest.

References

1. Kongor, J.E.; Hinneha, M.; Van deWalle, D.; Afoakwa, E.O.; Boeckx, P.; Dewettinck, K. Factors influencing quality variation in cocoa (*Theobroma cacao*) bean flavor profile—A review. *Food Res. Int.* **2016**, *82*, 44–52. [CrossRef]

2. Quiroz-Reyes, C.N.; Fogliano, V. Design cocoa processing towards healthy cocoa products: The role of phenolics and melanoidins. *J. Funct. Foods* **2018**, *45*, 480–490. [CrossRef]

3. Rawel, H.M.; Huschek, G.; Sagu, S.T.; Homann, T. Cocoa Bean Proteins-Characterization, Changes and Modifications due to Ripening and Post-Harvest Processing. *Nutrients* **2019**, *11*, 428. [CrossRef]

4. Sacchetti, G.; Ioannone, F.; De Gregorio, M.; Di Mattia, C.; Serafini, M.; Mastrocola, D. Non enzymatic browning during cocoa roasting as affected by processing time and temperature. *J. Food Eng.* **2016**, *169*, 44–52. [CrossRef]

5. Ioannone, F.; Di Mattia, C.D.; De Gregorio, M.; Sergi, M.; Serafini, M.; Sacchetti, G. Flavanols, proanthocyanidins and antioxidant activity changes during cocoa (*Theobroma cacao* L.) roasting as affected by temperature and time of processing. *Food Chem.* **2015**, *174*, 256–262. [CrossRef] [PubMed]

6. Oliviero, T.; Capuano, E.; Cammerer, B.; Fogliano, V. Influence of roasting on the antioxidant activity and HMF formation of a cocoa bean model systems. *J. Agric. Food Chem.* **2009**, *57*, 147–152. [CrossRef] [PubMed]

7. Zyzelewicz, D.; Oracz, J.; Sosnowska, D.; Nebesny, E. The influence of the roasting process conditions on the polyphenol content in cocoa beans, nibs and chocolates. *Food Res. Int.* **2016**, *89*, 918–929. [CrossRef]

8. Oracz, J.; Nebesny, E. Antioxidant properties of cocoa beans (*Theobroma cacao* L.): Influence of cultivar and roasting conditions. *Int. J. Food Prop.* **2016**, *19*, 1242–1258. [CrossRef]

9. Wang, H.Y.; Qian, H.; Yao, W.R. Melanoidins produced by the Maillard reaction: Structure and biological activity. *Food Chem.* **2011**, *128*, 573–584. [CrossRef]

10. Chen, K.; Zhao, J.; Shi, X.; Abdul, Q.; Jiang, Z. Characterization and Antioxidant Activity of Products Derived from Xylose–Bovine Casein Hydrolysate Maillard Reaction: Impact of Reaction Time. *Foods* **2019**, *8*, 242. [CrossRef]

11. Morales, F.J.; Somoza, V.; Fogliano, V. Physiological relevance of dietary melanoidins. *Amino Acids* **2012**, *42*, 1097–1109. [CrossRef] [PubMed]

12. De Marco, L.M.; Fischer, S.; Henle, T. High molecular weight coffee melanoidins are inhibitors for matrix metalloproteases. *J. Agric. Food Chem.* **2011**, *59*, 11417–11423. [CrossRef] [PubMed]

13. Bekedam, E.K.; Schols, H.A.; Van Boekel, M.A.; Smit, G. High molecular weight melanoidins from coffee brew. *J. Agric. Food Chem.* **2006**, *54*, 7658–7666. [CrossRef] [PubMed]

14. Moreira, A.S.; Nunes, F.M.; Simões, C.; Maciel, E.; Domingues, P.; Domingues, M.R.; Coimbra, M.A. Transglycosylation reactions, a main mechanism of phenolics incorporation in coffee melanoidins: Inhibition by Maillard reaction. *Food Chem.* **2017**, *227*, 422–431. [CrossRef]

15. Coelho, C.; Ribeiro, M.; Cruz, A.C.; Domingues, M.R.; Coimbra, M.A.; Bunzel, M.; Nunes, F.M. Nature of phenolic compounds in coffee melanoidins. *J. Agric. Food Chem.* **2014**, *62*, 7843–7853. [CrossRef]

16. Summa, C.; Raposo, F.C.; McCourt, J.; Lo Scalzo, R.; Wagner, K.H.; Elmadfa, I.; Anklam, E. Effect of roasting on the radical scavenging activity of cocoa beans. *Eur. Food Res. Technol.* **2006**, *222*, 368–375. [CrossRef]

17. Summa, C.; McCourt, J.; Cämmerer, B.; Fiala, A.; Probst, M.; Kun, S.; Anklam, E.; Wagner, K.H. Radical scavenging activity, anti-bacterial and mutagenic effects of cocoa bean Maillard reaction products with degree of roasting. *Mol. Nutr. Food Res.* **2008**, *52*, 342–351. [CrossRef]

18. Bellesia, A.; Tagliazucchi, D. Cocoa brew inhibits in vitro α-glucosidase activity: The role of polyphenols and high molecular weight compounds. *Food Res. Int.* **2014**, *63*, 439–445. [CrossRef]

19. Oracz, J.; Nebesny, E. Effect of roasting parameters on the physicochemical characteristics of high-molecular-weight Maillard reaction products isolated from cocoa beans of different *Theobroma cacao* L. groups. *Eur. Food Res. Technol.* **2019**, *245*, 118–128. [CrossRef]

20. Oracz, J.; Nebesny, E.; Zyzelewicz, D. Identification and quantification of free and bound phenolic compounds contained in the high-molecular weight melanoidin fractions derived from two different types of cocoa beans by UHPLC-DAD-ESI-HR-MSn. *Food Res. Int.* **2019**, *115*, 135–149. [CrossRef]

21. Niemenak, N.; Cilas, C.; Rohsius, C.; Bleiholder, H.; Meier, U.; Lieberei, R. Phenological growth stages of cacao plants (Theobroma sp.): Codification and description according to the BBCH scale. *Ann. Appl. Biol.* **2010**, *156*, 13–24. [CrossRef]

22. Belšcak, A.; Komes, D.; Horzic, D.; Kovacevic, K.; Karlovic, G.D. Comparative study of commercially available cocoa products in terms of their bioactive composition. *Food Res. Int.* **2009**, *42*, 707–716. [CrossRef]

23. Pastoriza, S.; Rufián-Henares, J.A. Contribution of melanoidins to the antioxidant capacity of the Spanish diet. *Food Chem.* **2014**, *164*, 438–445. [CrossRef] [PubMed]

24. Gu, F.L.; Kim, J.M.; Abbas, S.; Zhang, X.M.; Xia, S.Q.; Chen, Z.X. Structure and antioxidant activity of high molecular weight Maillard reaction products from casein–glucose. *Food Chem.* **2010**, *120*, 505–511. [CrossRef]

25. Batista, N.N.; de Andrade, D.P.; Ramos, C.L.; Dias, D.R.; Schwan, R.F. Antioxidant capacity of cocoa beans and chocolate assessed by FTIR. *Food Res. Int.* **2016**, *90*, 313–319. [CrossRef] [PubMed]

26. Smith, B.C. *Infrared Spectral Interpretation A Systematic Approach*, 1st ed.; CRC Press: Boca Raton, FL, USA; Washington, DC, USA, 1998; pp. 1–288. [CrossRef]

27. Pérez-Martínez, M.; Caemmerer, B.; Peña, M.P.; Cid, C.; Kroh, L.W. Influence of brewing method and acidity regulators on the antioxidant capacity of coffee brews. *J. Agric. Food Chem.* **2010**, *58*, 2958–2965. [CrossRef] [PubMed]

28. Abbès, F.; Kchaou, W.; Blecker, C.; Ongena, M.; Lognay, G.; Attia, H.; Besbes, S. Effect of processing conditions on phenolic compounds and antioxidant properties of date syrup. *Ind. Crops Prod.* **2013**, *44*, 634–642. [CrossRef]

29. Alves, G.; Perrone, D. Breads enriched with guava flour as a tool for studying the incorporation of phenolic compounds in bread melanoidins. *Food Chem.* **2015**, *185*, 65–74. [CrossRef]

30. Le Bourvellec, C.; Renard, C.M. Interactions between polyphenols and macromolecules: Quantification methods and mechanisms. *Crit. Rev. Food Sci. Nutr.* **2012**, *52*, 213–248. [CrossRef]

31. Nunes, F.M.; Coimbra, M.A. Melanoidins from coffee infusions. Fractionation, chemical characterization, and effect of the degree of roast. *J. Agric. Food Chem.* **2007**, *55*, 3967–3977. [CrossRef]

32. Li, X.; Chen, B.; Xie, H.; He, Y.; Zhong, D.; Chen, D. Antioxidant Structure–Activity Relationship Analysis of Five Dihydrochalcones. *Molecules* **2018**, *23*, 1162. [CrossRef] [PubMed]
33. Nooshkam, M.; Varidi, M.; Bashash, M. The Maillard reaction products as food-born antioxidant and antibrowning agents in model and real food systems. *Food Chem.* **2019**, *275*, 644–660. [CrossRef] [PubMed]
34. Giuffrè, A.M.; Zappia, C.; Capocasale, M. Physico-chemical Stability of Blood Orange Juice during Frozen Storage. *Int. J. Food Prop.* **2017**, *20*, 1930–1943. [CrossRef]
35. Giuffrè, A.M.; Zappia, C.; Capocasale, M. Effects of high temperatures and duration of heating on olive oil properties for food use and biodiesel production. *J. Am. Oil Chem. Soc.* **2017**, *6*, 819–830. [CrossRef]
36. Rameshwar Naidu, J.; Ismail, R.B.; Chen, Y.; Sasidharan, S.; Kumar, P. Chemical composition and antioxidant activity of the crude methanolic extracts of Mentha spicata. *J. Phytol.* **2012**, *4*, 13–18.
37. Shannon, E.; Jaiswal, A.K.; Abu-Ghannam, N. Polyphenolic content and antioxidant capacity of white, green, black, and herbal teas: A kinetic study. *Food Res.* **2018**, *2*, 1–11. [CrossRef]
38. Delgado-Andrade, C.; Rufián-Henares, J.A.; Morales, F.J. Assessing the antioxidant activity of melanoidins from coffee brews by different antioxidant methods. *J. Agric. Food Chem.* **2005**, *53*, 7832–7836. [CrossRef]
39. Ichikawa, K.; Sasada, R.; Chiba, K.; Gotoh, H. Effect of Side Chain Functional Groups on the DPPH Radical Scavenging Activity of Bisabolane-Type Phenols. *Antioxidants* **2019**, *8*, 65. [CrossRef]
40. Brudzynski, K.; Miotto, D. The recognition of high molecular weight melanoidins as the main components responsible for radical-scavenging capacity of unheated and heat-treated Canadian honeys. *Food Chem.* **2011**, *127*, 1023–1030. [CrossRef]
41. Tabart, J.; Kevers, C.; Pincemail, J.; Defraigne, J.O.; Dommes, J. Comparative antioxidant capacities of phenolic compounds measured by various tests. *Food Chem.* **2009**, *113*, 1226–1233. [CrossRef]
42. Mohsin, G.F.; Schmitt, F.J.; Kanzler, C.; Dirk Epping, J.; Flemig, S.; Hornemann, A. Structural characterization of melanoidin formed from d-glucose and l-alanine at different temperatures applying FTIR, NMR, EPR, and MALDI-ToF-MS. *Food Chem.* **2018**, *245*, 761–767. [CrossRef] [PubMed]
43. Minatel, I.O.; Borges, C.V.; Ferreira, M.I.; Gomez, A.G.; Chen, C.-Y.O.; Lima, G.P.P. Phenolic Compounds: Functional Properties, Impact of Processing and Bioavailability. In *Phenolic Compounds—Biological Activity*; Soto-Hernández, M., Palma-Tenango, M., García-Mateos, R., Eds.; InTech: London, UK, 2007; pp. 1–24. [CrossRef]
44. Zamora, R.; Hidalgo, F.J. The Maillard reaction and lipid oxidation. *Lipid Technol.* **2011**, *23*, 59–62. [CrossRef]
45. Oliveira, R.N.; Mancini, M.C.; Oliveira, F.C.; Passos, T.M.; Quilty, B.; Thiré, R.M.; McGuinness, G.B. FTIR analysis and quantification of phenols and flavonoids of five commercially available plants extracts used in wound healing. *Matéria (Rio J.)* **2016**, *21*, 767–779. [CrossRef]
46. Mot, A.; Silaghi-Dumitrescu, R.; Sârbu, C. Rapid and effective evaluation of the antioxidant capacity of propolis extracts using DPPH bleaching kinetic profiles, FT-IR and UV-vis spectroscopic data. *J. Food Composit. Anal.* **2011**, *24*, 516–522. [CrossRef]

Article

Antioxidant Effect of Cocoa By-Product and Cherry Polyphenol Extracts: A Comparative Study

Francesca Felice [1,*], Angela Fabiano [2], Marinella De Leo [2,3], Anna Maria Piras [2], Denise Beconcini [2], Maria Michela Cesare [1,4], Alessandra Braca [2,3], Ylenia Zambito [2,3] and Rossella Di Stefano [1,3]

[1]   Cardiovascular Research Laboratory, Department of Surgical, Medical and Molecular Pathology and Critical Care Medicine, University of Pisa, 56100 Pisa, Italy; maria.cesare@student.unisi.it (M.M.C.); rossella.distefano@unipi.it (R.D.S.)

[2]   Department of Pharmacy, University of Pisa, 56126 Pisa, Italy; angela.fabiano@unipi.it (A.F.); marinella.deleo@unipi.it (M.D.L.); anna.piras@unipi.it (A.M.P.); denisebeconcini@gmail.com (D.B.); alessandra.braca@unipi.it (A.B.); ylenia.zambito@unipi.it (Y.Z.)

[3]   Interdepartmental Research Center "Nutraceuticals and Food for Health", University of Pisa, 56100 Pisa, Italy

[4]   Department of Life Sciences, University of Siena, Siena 53100, Italy

*   Correspondence: francesca.felice@for.unipi.it; Tel.: +39-050-99-5577

Received: 17 December 2019; Accepted: 31 January 2020; Published: 3 February 2020

Abstract: Background: Recent studies have highlighted the importance of cherry and cocoa extracts consumption to protect cells from oxidative stress, paying particular attention to cocoa by-products. This study aims to investigate the protective effect of cocoa husk extract (CHE) and cherry extracts (CE) against ROS-induced oxidative stress in Human Umbilical Vein Endothelial Cells (HUVECs). Methods: CE and CHE had antioxidant activity characterized by total polyphenols content (TPC). HUVECs were treated for 2 h and 24 h with increasing TPC concentrations of CE and CHE (5-10-25-50-100 µg Gallic Acid Equivalent (GAE)/mL) and then with H_2O_2 for 1 h. Cell viability and ROS production were evaluated. CE and CHE polyphenols permeability on excised rat intestine were also studied. Results: CE and CHE showed a similar antioxidant activity (2.5 ± 0.01 mmol Fe^{2+}/100 g FW (fresh weight) and 2.19 ± 0.09 mmol Fe^{2+}/100 g FW, respectively, $p > 0.05$) whereas CHE had a higher TPC (7105.0 ± 96.9 mg GAE/100 g FW) than CE (402.5 ± 8.4 mg GAE/100 g), $p < 0.05$. The in vitro viability assay showed that both extracts were non-cytotoxic. CHE resulted in protection against ROS at lower concentrations than CE. CHE showed a 2-fold higher apparent permeability compared to CE. Conclusions: CHE represents a high-value antioxidant source, which is interesting for the food and pharmaceutical industries.

Keywords: cocoa by-products; cherry extract; oxidative stress; human endothelial cell

---

1. Introduction

Reactive oxygen species (ROS) are involved in the pathogenesis of numerous chronic and degenerative diseases. Physiologically, oxygen metabolism generates ROS, which are contrasted and neutralized by antioxidant defenses. The unbalance toward ROS formation is recognized as a critical aspect of cell damage that characterizes many disease states, such as atherosclerosis and premature aging [1–3].

Vascular endothelial cell lines are particularly sensitive to ROS, and damage to them is reflected in the alteration of vascular tone and permeability, and thus involved in cardiovascular dysfunction associated with hypertension, diabetes, atherosclerosis, and ischemic heart disease [4–6].

Therefore, the use of nutraceuticals appears essential for the prevention and control of ROS induced damages. In nature, polyphenols are the most abundant category of antioxidants. Most of them are in fruits and vegetables and are typically associated with healthy diets. Actually, epidemiological studies and associated meta-analyses have confirmed that their long-term consumption correlates with protection against the development of serious diseases such as cancers, cardiovascular diseases (CVD), diabetes, osteoporosis, and neurodegenerative pathologies [7].

Cocoa and its derivative products are rich in polyphenols, which possess an antioxidant capacity and are associated with the prevention of diseases related to oxidative stress. Most of the phenolic compounds found in cocoa are represented by flavonols, such as catechins, or flavan-3-ols (37%), anthocyanins (4%), and proanthocyanidins (58%) [8,9]. The protective activity of cocoa seems to be due to its phytochemical constituents, especially catechins.

Cocoa by-products, mainly cocoa husk, are produced worldwide in large amounts, constituting about 75% wt of whole fruit from the cocoa harvest [10]. Being a production waste, they are generally discarded, with a negative environmental impact [11]. Moreover, the cocoa by-products (bean and husks) possess nutritional and functional properties, particularly related to the presence of proanthocyanidins identified in the husks [12]. The proanthocyanidins found in husk are tannins, which can have different molecular weights, according to the degree of polymerization [13]. In vitro cellular studies have also confirmed that proanthocyanidins possess antioxidant activity [8,14]. Additionally, flavanols protect cells from oxidative stress by reducing ROS production and inhibiting the activation of caspase-3. Procyanidin B2 also increases the performance of enzymes specifically involved in antioxidant and detoxification processes [15].

Along with cocoa, cherry fruits have valuable nutritional properties, and their beneficial effects have been demonstrated against oxidative stress damage on both neuronal and intestinal epithelial cells. Similarly to cocoa, the most representative nutraceutical actives in cherries are polyphenols, including flavonoids and anthocyanins [16]. Many studies have used natural extracts on endothelial progenitor cells [17,18] or Human Umbilical Vein Endothelial Cells (HUVECs) [19–21] in in vitro experiments related to vascular dysfunction.

The aim of the present research was to compare the antioxidant properties of two antioxidant-rich natural products plant extracts, namely cherry and cocoa (bean and husk). The extracts were evaluated both in vitro and ex vivo in order to compare their beneficial effects on vascular related dysfunction upon oral intake.

For this purpose, we performed phenol cocoa bean and husk extraction from two different plant varieties (Costa Rica and Madagascar) that had their antioxidant activity, total polyphenols content, and phenolic composition characterized by HPLC coupled to electrospray ionization tandem mass spectrometry (ESI-MS/MS). The characterization of cherry extract (CE) has already been reported [22]. Moreover, Costa Rica cocoa husk extract (CHE) and CE were compared for their ability to permeate across excised rat intestine.

2. Materials and Methods

2.1. Materials

Hexane, acetone, and Folin-Ciocalteau reagent, and gelatin were purchased from Sigma-Aldrich (Milan, Italy). HPLC grade formic acid and methanol were purchased from VWR (Milan, Italy). HPLC grade water (18 mΩ) was obtained by a Mill-Q purification system (Millipore Corp., Bedford, MA, USA).

H_2O_2 was purchased from Farmac-Zabban S.p.a. (Calderara di Reno, BO, Italy). Medium 199 (M199), fetal bovine serum (FBS), penicillin-streptomycin solution, L-glutamine, and HEPES buffer were supplied by Hospira S.r.l. (Naples, Italy).

4-[3-(4-iodophenyl)-2-(4-nitrophenyl)-2H-5-tetrazolium]-1,3-benzenedisulfonate (WST-1 assay), was purchased from Roche Applied Science (Mannheim, Germany), 5-(and-6)-chloromethyl-2′,

7′-dichloro-di-hydro-fluorescein diacetate, and acetyl ester (CM – H_2DCFDA) were supplied by Thermo Fisher Scientific Inc. (Waltham, MA, USA).

2.2. Sample Preparation

2.2.1. Cocoa Bean and Husk Phenol Extraction

The cocoa bean and husk phenol extraction were carried out using a procedure reported by Hammerstone et al. [23], slightly modified. Two varieties of cocoa (Costa Rica and Madagascar) were ground for 30 s in order to obtain a homogeneous material that was defatted four times with 125 mL of hexane for 20 min at 200 rpm and subsequently centrifuged for 30 min at 4000 rpm. The defatted cocoa sample was then extracted four times with acetone 70% *v/v* at a ratio of 1:5, stirred for 3 min, and centrifuged for 30 min at 4000 rpm. The extraction procedure was repeated four times after which the supernatants were combined, filtered with a paper filter and the organic solvent was removed by evaporation at room temperature for 48 h. To obtain a stable product the remaining water suspension was lyophilized.

The cocoa from Madagascar was selected because Bruna et al. [24], comparing the content of polyphenols in cocoa husks from different countries (Ghana, Madagascar, Ecuador, Trinidad, and Venezuela), found that Madagascar husks were the richest. The cocoa from Costa Rica was chosen because compared to that from other countries (Ivory Coast, Venezuela, Samoa, Trinidad, Brazil, Ghana, Ecuador, Jamaica), it was found to be particularly rich in epicatechin [25–28].

2.2.2. Cherry Extract (CE) Preparation and Characterization

The cherry extracts were obtained from the Crognola Capannile variety of *Prunus avium* L., an ancient Tuscan variety of sweet cherry, as described by Beconcini et al. [29].

The Crognola Capannile variety was used for the present study due to their high polyphenol content, as reported by Berni et al. [22].

2.3. HPLC-PDA/UVvis-ESI-MS/MS Analysis of Cocoa Extracts

Cocoa extracts were dissolved in methanol at a final concentration of 2.5 mg/mL, then centrifuged for 5 min at 1145× *g*. The supernatants (20 μL injection volume) were subjected to HPLC coupled with a photodiode array (PDA)/UVvis and an ion trap ESI-MS. The LC system was composed of a Surveyor autosampler, a Surveyor LC pump, a Surveyor PDA/UVvis detector, and a LCQ Advantage ion trap ESI mass spectrometer (ThermoFinnigan, San Jose, CA, USA) equipped with Xcalibur 3.1 software. Qualitative analysis was performed on a C-18 column (Luna, 4.6 × 150 mm, 5 μm, Phenomenex, Bologna, Italy) using a mixture of methanol (solvent A) and formic acid in water 0.1% *v/v* (solvent B) as eluent. A linear gradient of increasing 5 to 55% A was developed within 50 min at a flow rate of 0.8 mL/min, with a splitting system of 2:8 to MS detector (160 μL/min) and PDA detector (640 μL/min), respectively. PDA/UV spectra were recorded in a range of 200–600 nm, using 254, 280, and 325 nm as preferential channels, while ESI-MS experiments were achieved in negative ion mode (scan range of *m/z* 150-2000). Ionization parameters were used as previously reported [30].

2.4. Antioxidants Determination

The total antioxidant potential of cocoa by-product extracts and CE freeze-dried samples was determined by a ferric reducing antioxidant power (FRAP) assay, as previously reported [31]. The FRAP value of the samples, expressed as mmol of Fe^{2+} per 100 g FW, was determined from a standards curve built up using ferrous sulfate.

2.5. Total Polyphenolic Content

The total polyphenols content (TPC) of cocoa by-product extract and CE was determined by the Folin–Ciocalteau method [32]. The results were expressed as gallic acid equivalent (GAE) on a dry weight basis following a previously reported procedure [33].

2.6. HUVEC Isolation and Culture

HUVECs were collected from umbilical cords obtained from healthy donors women post-labor, treated anonymously. The age of the donors ranged from 24 to 43 years. Ethical approval (authorization number 2943), was granted by the local Ethics Committee (Full name: Comitato Etico per la Sperimentazione Clinica Area Vasta Nord Ovest c/o Azienda Ospedaliero-Universitaria Pisana (AOUP), Pisa"). HUVECs were collected following the procedure described by Jeffe et al. [34]. Briefly, HUVECs after isolation and centrifugation were plated on gelatin pre-coated flasks and incubated for 24 h at 37 °C, using 5% CO_2 in a complete growth medium made with 10% FBS. After 24 h, the growth medium was replaced to remove the excess of red blood cells.

2.7. Cell Treatment

HUVECs between passage P2–P4 were treated for 2 h and 24 h with increasing polyphenol concentrations of cocoa husk Costa Rica extract (CHE) and Crognola Capannile cherry extract (CE) (5, 10, 25 or 50 µg GAE/mL), in growth medium with 5% FBS. Cells in medium only were used as a positive control. Then, cells were washed with PBS and treated with 50 µM of commercial H_2O_2 for 1 h to induce oxidative stress [33]. At the end of each treatment, cells were analyzed for viability and ROS production.

2.8. Cell Viability

At the end of each treatment, HUVECs were incubated with tetrazolium salt (10 µL/well) for 3 h at 37 °C, in 5% CO_2 and the formazan dye formed was quantified at 450 nm with a multiplate reader (Thermo Scientific Multiskan FC Microplate Photometer), correlating the with the number of active cells. The viability was expressed as the percentage of viable cells.

2.9. ROS Production

ROS production was evaluated using CM – H2DCFDA, a fluorescent probe. HUVECs during the last 15 min of treatment with CHE, CE or H_2O_2, were incubated in the dark at room temperature with CM – H2DCFDA (10 µL/well) dissolved in PBS. ROS production was detected measuring the increase in fluorescence over time at excitation of 488 nm and emission of 510 nm using a microplate reader (Thermo Scientific Fluoroskan Ascent Microplate Fluorometer). ROS production was expressed as a percentage of ROS accumulation.

2.10. Permeation Study of CHE and CE

For the ex-vivo permeation studies, we used the intestinal mucosa excised from non-fasting male Wistar rats (weight 250–300 g), using the procedure reported by Fabiano et al. [35]. The ex vivo experiments on excised rat intestine were carried out with the aim of assessing the permeability of the intestinal epithelium to the antioxidants present in the different extracts. The isolated rat intestine was chosen from among the known intestinal models because its tight junctions are similar in number and tightness to those of the human intestine [36]. The formulation tested were 1 mL of CE solution (15.5 µg/mL GAE concentration [33], or CHE (280 µg/mL GAE concentration). At 30 min intervals of a total of 240 min, the apical to basolateral transport of CHE or CE was investigated, analyzing the receiving phase (50 µL) by the Folin–Ciocalteau reagent for TPC.

2.11. Cocoa Extract and CE Stability Studies

The stability of CHE and CE was evaluated according to the procedure reported by Beconcini et al. [33]. At 30 min intervals of a total of 240 min, 50 µL of CHE or CE volume was withdrawn and analyzed by the Folin–Ciocalteau reagent for TPC.

2.12. Statistical Analysis

The GraphPad Prism Software vs. 7.0 (GraphPad Software Inc., La Jolla, CA, USA) was used for the statistical analysis of data. All results were presented as means ± standard deviation (SD) of at least three independent experiments. The significant difference (p-value < 0.05) between groups of values was evaluated by a one-way ANOVA or Turkey's or Bonferroni's multiple comparisons.

3. Results

3.1. Phenolic Profile of Cocoa Extracts

The phenolic composition of cocoa bean and husk extracts was characterized by HPLC-PDA/UVvis-ESI-MS/MS experiments. Cocoa Costa Rica extracts (bean and husk) showed similar chemical profiles (Figure 1), with 14 major phenol compounds identified: one caffeoyl derivative, caffeoyl aspartate (**1**); one flavan-3-ol, catechin/epicatechin (**9**); three procyanidin B isomers (**5**, **6**, and **10**) together with seven procianydin C isomers (**2**, **3**, **4**, **7**, **8**, **11**, and **12**); two flavonol glycosides, quercetin 3-O-glucoside (**13**), and quercetin 3-O-arabinoside (**14**).

In contrast, the Madagascar cocoa husk extract showed a less rich phenolic content, mainly with regard to procyanidins. Indeed, compounds **5-9**, which were the most representative constituents in cocoa husk Costa Rica extract, were significantly reduced in Madagascar cocoa husk, while compounds **10**, **11**, and **12** were not detected at all. These findings are in agreement with previous studies, reporting a high different phenol content in cocoa products with different geographic origins, i.e., Costa Rica and Madagascar [24].

All compounds were tentatively identified by comparison of their elution order, UV data, and both full and fragmentation mass spectra (Table 1) with data reported in the literature [13,37,38]. Compounds **2–4**, **7**, **8**, **11**, and **12** showed the same molecular deprotonated ion [M − H]$^-$ at m/z 865 and two strong UV absorptions at 242–258 and 277–280 nm, suggesting a trimeric B-type procyanidin structure for all isomers. This finding was confirmed by fragmentation mass spectra, all showing the same diagnostic product ions at m/z 739, 713, 695, 451, 407, and 287, together with a base peak ion at m/z 577, represented by the dimeric form [30].

Since all the spectra are superimposable, it is not possible to distinguish between the isomeric forms [39], but all compounds can be assigned as procyanidin C isomers, previously found in cocoa extract [37]. Likewise, full MS of peaks **5**, **6**, and **10** were characterized by type-B procyanidin dimers, as deduced by the deprotonated molecule [M − H]$^-$ at m/z 577 and product ions at m/z 451, 425, 407, and 289. Also, in this case, the exact structure cannot be assigned only on the basis of spectral data. However, it can be assumed all three molecules to be isomers of procyanidin B, previously reported in cocoa extract [37]. In addition to these oligomers, also their monomer was detected (compound **9**, [M − H]$^-$ at m/z 289), corresponding to catechin or epicatechin, as deduced by diagnostic product ions at m/z 245 and 205 [30].

Along with flavan-3-ols, also two flavonol glycosides were identified in all three analyzed cocoa extracts, as evidenced by UV absorptions at 267–268 and 354–355 nm. Compounds **13** ([M − H]$^-$ at m/z 463) and **14** ([M − H]$^-$ at m/z 433) were quercetin derivatives showing the loss of a hexose ([M − H-162]$^-$ at m/z 301) and a pentose units ([M − H-132]$^-$ at m/z 301), respectively. According to data reported in previous work [40], **13** and **14** were identified as quercetin 3-O-glucoside and quercetin 3-O-arabinoside, respectively.

Finally, a caffeoyl derivative (compound **1**, [M − H]$^-$ at m/z 293) was detected in all cocoa samples, showing the loss of an aspartate residue ([M − H-115]$^-$ at m/z 301) in the ESI-MS/MS experiments.

Thus, compound **1** was identified as *N*-caffeoyl aspartate, previously reported in cocoa source [37]. Some other peaks remained unidentified; however, based on UV data, they were not attributed to phenol derivatives.

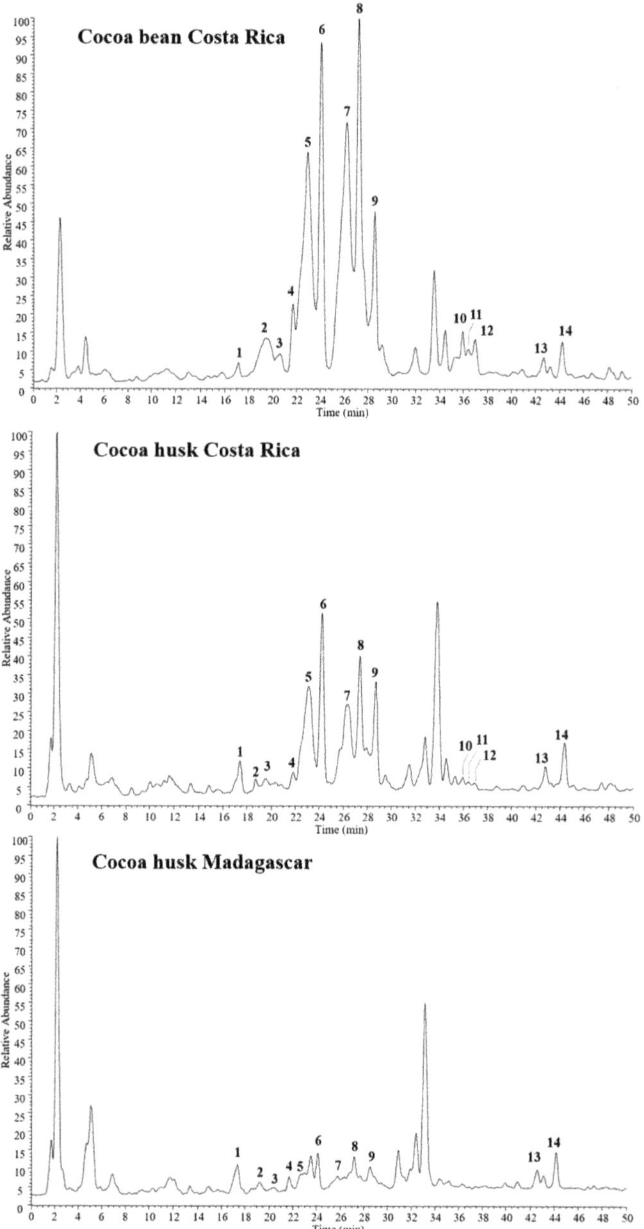

Figure 1. HPLC-electrospray ionization (ESI)-MS/MS profiles of phenols detected in cocoa husk and bean extracts in negative ion mode. Peak data are listed in Table 1.

Table 1. ESI-MS/MS, UV, and chromatographic data (retention time, t_R) of compounds **1–14** detected in the cocoa bean and husk extracts Costa Rica and cocoa husk Madagascar.

Peak [a]	Compound	t_R (min)	M	[M+HCOO]⁻	[M−H]⁻	ESI MS/MS (Product Ions) (m/z) [b]	UV (λ_{max})
	Phenols						
1	*N*-caffeoyl aspartate	17.8	295		294	276, **179**, 132	252, 277, 305
2	procyanidin C (trimer I)	19.4	866		865	847, 739, 713, 695, 577, 451, 407, 287	258, 277
3	procyanidin C (trimer II)	20.7	866		865	847, 739, 713, **695**, 577, 451, 407, 287	252, 279
4	procyanidin C (trimer III)	21.7	866		865	847, 739, 713, **695**, 577, 451, 407, 287	248, 280
5	procyanidin B (dimer I)	23.0	578		577	451, **425**, 407, 289	243, 279
6	procyanidin B (dimer II)	24.1	578		577	451, **425**, 407, 289	243, 279
7	procyanidin C (trimer IV)	26.3	866		865	847, 739, 713, **695**, 577, 451, 407, 287	244, 280
8	procyanidin C (trimer V)	27.2	866		865	847, 739, 713, **695**, 577, 451, 407, 287	242, 279
9	catechin/epicatechin	28.6	290	335	289	271, **245**, 205, 179	240, 279
10	procyanidin B (dimer III)	36.0	578		577	451, **425**, 407, 289	277
11	procyanidin C (trimer VI)	36.4	866		865	739, 713, **695**, 577, 451, 407, 287	277
12	procyanidin C (trimer VII)	37.0	866		865	739, 713, **695**, 577, 451, 407, 287	278
13	quercetin 3-*O*-glucoside	42.7	464		463	**301**, 179	268, 355
14	quercetin 3-*O*-arabinoside	44.2	434		433	**301**, 179	267, 354

[a] Peak numbers correspond with those of Figure 1. [b] Ions were generated by fragmentation of molecular deprotonated ions in the ESI-MS/MS experiments, and the base peaks are showed in bold.

3.2. Cherry and Cocoa By-Product Extracts Characterization

FRAP and Folin–Ciocalteu methods on CE reported that Crognola had the highest antioxidant content and TPC (402.5 ± 8.4 mg GAE/100 g FW) among the six varieties of Prunus avium L. studied. Flavonoid molecules of quercetin and catechins were 59.32 ± 3.2 µg/g FW and 292.76 ± 1.9 µg/g FW, respectively. HPLC analysis also showed the presence of anthocyanins, represented mainly by cyanidin-3-glucoside (227.37 ± 1.2 µg/g FW).

In Table 2, the antioxidant content and the TPC of the two extracts under study are reported. CE and CHE antioxidant content was not significantly different ($p > 0.05$), whereas CHE had a higher TPC than CE ($p < 0.05$).

Table 2. Cherry Extract (CE) and Cocoa Husk Extract (CHE) characterization. [a] Determined by FRAP. [b] Determined by Folin–Ciocalteau. * Significantly different from each other ($p < 0.05$). TPC = total polyphenols content.

Extract	Antioxidant Content [a] (mmol Fe²⁺/100 g FW)	TPC [b] (mg GAE/100g FW)
CE	2.19 ± 0.09	402.5 ± 8.4
CHE	2.50 ± 0.01	7105 ± 96.9 *

3.3. Dose- and Time-Dependent Effect of CHE and CE on HUVECs Viability

Polyphenolic concentrations of 5, 10, 25, 50 and 100 µg GAE/mL, were chosen for viability studies. Cell viability was evaluated by WST-1 colorimetric assay. Both CHE and CE polyphenols were non-cytotoxic, both after 2 h and 24 h of treatment (Figure 2a,b). After 24 h of treatment, there was an increase in cell viability at high concentrations of CHE (25 to 100 µg GAE/mL TPC) compared to control (Figure 2b). This enhancement of cell viability might be due to the incubation over time addicted to a higher concentration of polyphenols, as previously demonstrated in apple juice study on HUVECs [20].

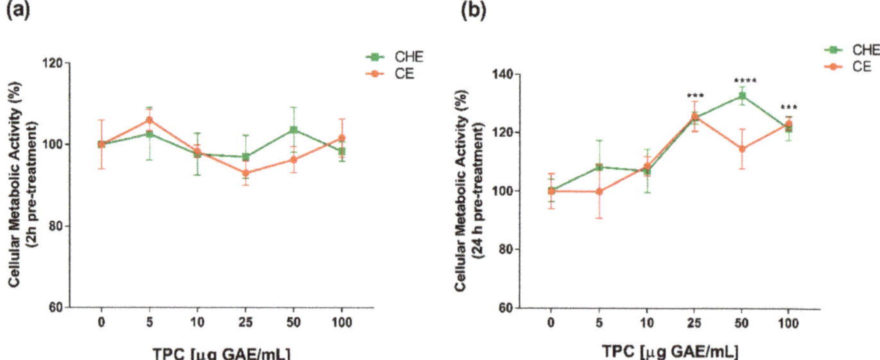

Figure 2. Dose- and time-dependent cell metabolic activity. Human Umbilical Vein Endothelial Cells (HUVECs) were cultured for 2 h (**a**) and 24 h (**b**) in the presence of increasing concentrations of total polyphenol content (TPC) from CE or CHE (5, 10, 25, 50, and 100 µg GAE/mL). Cell metabolic activity was determined by WST-1 colorimetric assay and expressed as metabolic activity percentage compared to control (untreated cells). Graphical data are represented as mean ± SD of three separate experiments run in triplicate. (*** $p < 0.005$, **** $p < 0.0001$ vs. control).

3.4. Protective Effect from Oxidative Stress

To evaluate the antioxidant activity of CHE and CE, HUVECs were pre-treated for 2 h and 24 h with increasing polyphenols concentrations (ranging from 5–100 µg GAE/mL of TPC), and after that, an oxidative stress insult was applied by treating the cells with 50 µM H_2O_2 for 1 h.

The results reported that treatment of HUVECs with H_2O_2 significantly reduced viable cell number compared to control (Figure 3). Viability after H_2O_2-induced oxidative stress was increased by a pre-treatment with CE, both after 2 h or 24 h (Figure 3b). In particular, after 24 h, only 50 µg GAE/mL TPC of CE mL had the most significant protective effect. For CHE polyphenols, only the 24 h pre-treatment significantly protected cells from oxidative stress in a concentration range of 5 µg GAE/mL TPC to 50 µg GAE/mL TPC.

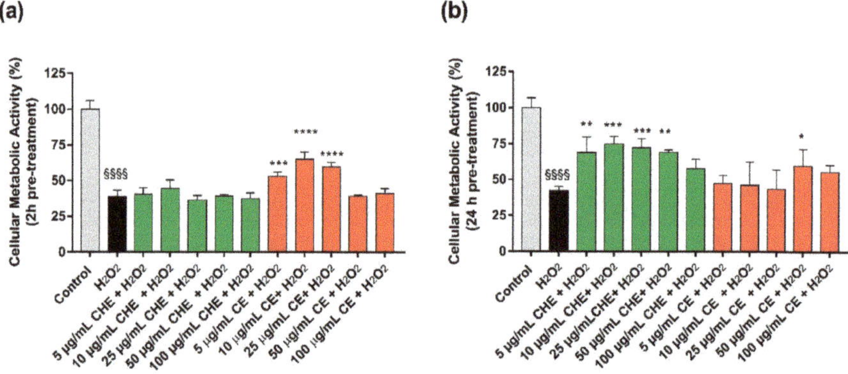

Figure 3. Protective effects from H_2O_2-induced oxidative stress. HUVECs viability after 2 h (**a**) or 24 h (**b**) of pre-treatment with CE or CHE and treatment with 50 µM H_2O_2 for 1 h. Data are expressed as the % of viable cells compared to 100% of control (untreated cells). (*$p < 0.05$, ** $p < 0.005$, *** $p < 0.0005$, **** $p < 0.0001$ vs. H_2O_2; §§§§ $p < 0.0001$ vs. control).

3.5. Antioxidant Activity

ROS accumulation in HUVECs was evaluated after 2 h and 24 h pre-treatment of CHE or CE at different concentrations (ranging 5–50 µg GAE/mL of TPC). Treatment of HUVECs with H_2O_2 significantly increased intracellular ROS production. As shown in Figure 4a, after 2 h pre-incubation, CHE showed to significantly reduce ROS production already at low concentration (10 µg GAE/mL TPC) compared with control, while a significantly protective effect by CE polyphenols was observed at a concentration above 25 µg GAE/mL (** $p < 0.005$ vs. control).

CHE and CE seemed to increase their antioxidant power over incubation time, as observed in Figure 4b, and directly correlated to their polyphenolic content, indicating a time- and dose-dependent effect on cell metabolic activity.

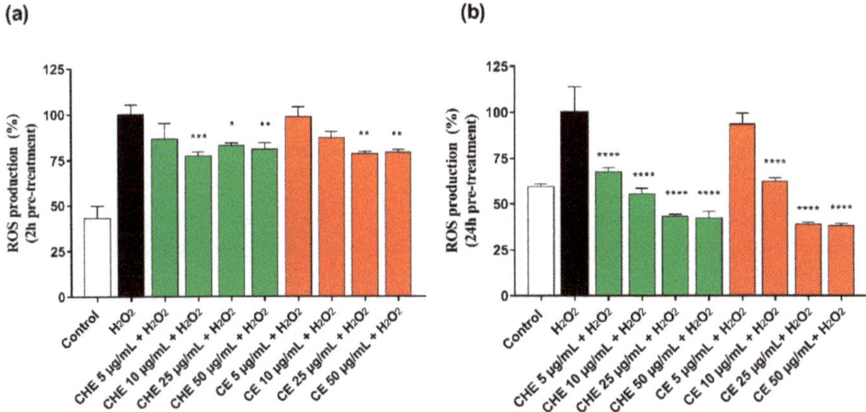

Figure 4. Rective Oxygen Species (ROS) production by HUVECs was evaluated after 2 h (**a**) and 24 h (**b**) of incubation with different concentrations of CHE and CE (i.e., 5, 10, 25, 50, and 100 µg/mL of TPC) and 100 µM H_2O_2 for 1 h. Data are expressed as ROS production% by treated cells and are representative of three separate experiments run in triplicate (* $p < 0.05$, ** $p < 0.005$, *** $p < 0.0005$, **** $p < 0.0001$ vs. H_2O_2).

3.6. Permeation Study of CHE and CE

For each permeation run, the value of the apparent permeability coefficient, P'$_{app}$, for permeant across the excised rat intestinal mucosa was calculated from the following equation

$$P'_{app} = dM/dt \times 1/(AC_0) \tag{1}$$

where dM/dt 1/A, the permeation flux, is the slope of the linear portion of the cumulative amount permeated per unit surface area vs. time plot, and C_0 is the CHE or CE concentration (280 µg/mL or 15.5 µg/mL, respectively) introduced at the donor phase. For each plot, the quality of the linear regression, shown in Figure 5, as evidenced by r^2 values being >0.9.

Such linearity allowed the application of Equation (1), as reported by [41]. The single P'$_{app}$ values were averaged to calculate the mean apparent permeability P$_{app}$ (n = 3). Since the extract concentration was different for CHE and CE, we calculated the cumulative percentage permeated at each time, useful for comparative purposes. The relevant data reported in Table 3 shows that for cocoa extract, the apparent permeability was 2-fold higher than that corresponding to CE, as the flux was 36-fold higher than that of CE.

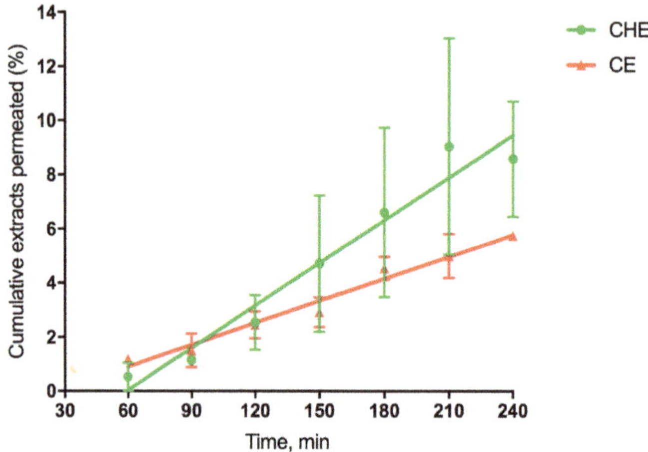

Figure 5. Cumulative CHE and CE permeated (%) across excised rat intestine.

Table 3. Data on CE or cocoa extract permeation across the excised rat intestine. [a] Apparent permeability. [b] Cumulative extract permeated (%) over the whole experiment time (4 h). * Significantly different from each other ($p < 0.05$).

Extract	Flux 10^2 (μg cm^{-2}min^{-1})	P_{app} [a] 10^4 (cm min^{-1})	T_{4h} [b] (%)
CE	0.41 ± 0.03	2.64 ± 0.02	5.75 ± 0.07
CHE	14.50 ± 1.33 *	5.18 ± 0.47 *	8.58 ± 2.13

3.7. CHE and CE Stability Studies

CHE and CE stability in simulated gastric fluids (SGFs) shown in Figure 6, indicates that CHE is more stable than CE in the stomach for at least 4 h. In fact, the CHE polyphenols percentage in the SGF was around 100% for the entire duration of the experiment, whereas that corresponding to CE decreased around 50% just after 3 h.

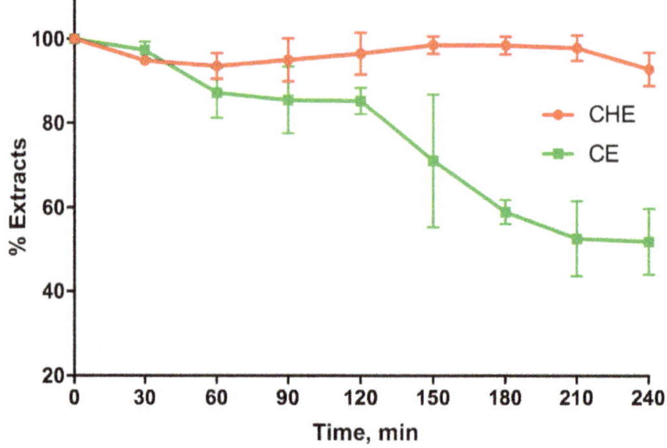

Figure 6. Cocoa or CE stability in simulated gastric fluid.

4. Discussion

Natural products are increasingly used in scientific research for their potential effect on human health. Many studies show their beneficial effects in clinical and in vitro studies due to the presence of antioxidant molecules, mainly polyphenols. Polyphenols have shown their effects against chronic diseases, including CVD. For the treatment of CVD, prevention plays an important role. The introduction of nutraceuticals in the diet could represent the first defense mechanism of the body from oxidative stress.

In the present study, we compare the antioxidant effects of the molecules contained in an ancient variety of Tuscan cherries (Crognola Capannile) and cocoa by-product extracts (bean and husk).

Numerous studies reported that sweet cherries had high and variable concentrations of antioxidants and TPC in different varieties. A recent study by Berni et al., found the same differences in Tuscan cherries varieties [22]. The analysis confirmed the presence of a high content of antioxidant and phenolic compounds, particularly in the variety Crognola Capannile for CE and Costa Rica for cocoa by-products extract, mainly husks. Specifically, the characterization of the phenolic composition of cocoa bean and husk extracts demonstrated a high concentration, mainly with regard to procyanidins. Above all, compounds **5–9** were the most representative constituents in cocoa husk Costa Rica extract. According to previous studies, cocoa extracts were found to be rich in flavonoids, in particular, dimeric and trimeric procyanidins, among which, the B-type (characterized by a C4–C8 or C4–C6 bond between monomers catechin/epicatechin) were the most representative [37].

Several studies have tested natural products-derived polyphenols on HUVECs [19–21] for in vitro experiments related to vascular dysfunction. In this study, we have investigated the in vitro properties of Crognola CE, obtained from fresh fruits, and selected from the six Tuscan varieties the highest TPC and molecular content (as previously shown by Berni et al. [22]), and Costa Rica cocoa by-product extract (in particular, husk product (CHE)), for their rich phenolic content [25].

Vascular oxidative stress contributes to mechanisms of vascular dysfunction and has been implicated in playing an important role in a number of cardiovascular pathologies [42]. There are also many in vitro studies about the antioxidant properties of *Prunus avium* L. and cocoa by-products for the prevention of chronic diseases [12,16,43,44]. The presence of hydroxycinnamic acids, flavonoids and anthocyanins made CE interesting for in vitro experiments related to oxidative stress. Matias et al. [16] demonstrated that a 2 h pre-treatment with cherry extracts effectively alleviated oxidative stress caused by H_2O_2-induced injury in neuronal cells.

In a recent study, Rebollo–Hernanz et al. [44] demonstrated that cocoa by-product extracts effectively reduced inflammatory markers in macrophages and adipocytes and the production of reactive oxygen species. Moreover, extracts from cocoa by-products also modulated the phosphorylation of the insulin receptor signaling pathway and stimulated GLUT-4 translocation, increasing glucose uptake [44].

Our results show that lower concentrations of CE protected from oxidative stress in a shorter treatment time (2 h, Figure 3a), in agreement with other studies [16], whereas a higher concentration of polyphenols was required in long-term treatment (24 h, Figure 3b). These results demonstrate the indirect antioxidant potential of CHE, possibly acting through the augmentation of cellular antioxidant capacity by enhancing specific genes encoding antioxidant proteins [45].

Polyphenols showed an anti-radical activity supported by different studies [12,46,47]. In this study, the antioxidant activity of CHE and CE polyphenolic molecules has been verified through the evaluation of ROS production, both with and without H_2O_2-stress induction on HUVECs.

The obtained results suggest that polyphenols in sweet CE and CHE are able to inhibit ROS, protecting cells from oxidative stress. In particular, CHE showed a time- and dose-dependent effect on cell metabolic activity, probably due to an indirect antioxidant effect [45]. Specifically, CHE showed an antioxidant effect at low concentrations.

Finally, in the present study, we evaluated the ability of natural extracts to cross the excised intestinal wall. The results indicate that CHE is more able to permeate through the excised intestinal wall compared to CE, probably because the antioxidants contained in CHE are more stable than those in CE when they come into contact with the rat intestinal tissue. This hypothesis is in agreement with the results shown in Figure 6, indicating that CHE is more stable than CE in the gastric environment.

5. Conclusions

The two extracts studied in this work, CHE and CE, have both been shown to have antioxidant activity on HUVECs, i.e., on the cells of the endothelium of blood vessels, thus proving to be potential compounds for the prevention of CVD. Between the two extracts, CHE showed better performance on HUVECs, and greater permeability across the rat intestine than CE, perhaps due to its greater stability in the physiological environment. The results obtained in this work encourage us to continue the studies on cocoa husks extracts, which represent a waste product of cocoa processing and a cost for their disposal. Thanks to their beneficial properties, cocoa by-products might become a nutraceutical research topic for possible medical applications in CVD, as well as a potential ingredient for the food and pharmaceutical industries.

Author Contributions: Conceptualization: F.F., Y.Z. and R.D.S.; Data curation: A.F., M.D.L., and D.B.; Formal analysis: D.B.; Investigation, F.F., A.F., D.B., and M.M.C.; Methodology, F.F., A.F., M.D.L., A.M.P., D.B., and A.B.; Project administration, A.B., Y.Z., and R.D.S.; Supervision, A.B., Y.Z., and R.D.S.; Validation, F.F., A.F., M.D.L., and A.M.P.; Writing—original draft, F.F., A.F., and D.B.; Writing—review and editing, F.F., A.F., M.D.L., A.M.P., and R.D.S. All authors have read and agree to the published version of the manuscript.

Funding: This research received no external funding.

References

1. Chakraborti, T.; Ghosh, S.K.; Michael, J.R.; Batabyal, S.K.; Chakraborti, S. Targets of oxidative stress in cardiovascular system. *Mol. Cell. Biochem.* **1998**, *187*, 1–10. [CrossRef] [PubMed]
2. Sastre, J.; Pallardo, F.V.; Garcia de la Asuncion, J.; Vina, J. Mitochondria, oxidative stress and aging. *Free Radic. Res.* **2000**, *32*, 189–198. [CrossRef] [PubMed]
3. Stadtman, E.R.; Berlett, B.S. Reactive oxygen-mediated protein oxidation in aging and disease. *Drug Metab. Rev.* **1998**, *30*, 225–243. [CrossRef] [PubMed]
4. Landmesser, U.; Harrison, D.G. Oxidative stress and vascular damage in hypertension. *Coron. Artery Dis.* **2001**, *12*, 455–461. [CrossRef]
5. Zalba, G.; San Jose, G.; Moreno, M.U.; Fortuno, M.A.; Fortuno, A.; Beaumont, F.J.; Diez, J. Oxidative stress in arterial hypertension: Role of NAD(P)H oxidase. *Hypertension* **2001**, *38*, 1395–1399. [CrossRef] [PubMed]
6. Incalza, M.A.; D'Oria, R.; Natalicchio, A.; Perrini, S.; Laviola, L.; Giorgino, F. Oxidative stress and reactive oxygen species in endothelial dysfunction associated with cardiovascular and metabolic diseases. *Vasc. Pharmacol.* **2018**, *100*, 1–19. [CrossRef]
7. Pandey, K.B.; Rizvi, S.I. Plant polyphenols as dietary antioxidants in human health and disease. *Oxidative Med. Cell. Longev.* **2009**, *2*, 270–278. [CrossRef]
8. Miller, K.B.; Stuart, D.A.; Smith, N.L.; Lee, C.Y.; McHale, N.L.; Flanagan, J.A.; Ou, B.; Hurst, W.J. Antioxidant activity and polyphenol and procyanidin contents of selected commercially available cocoa-containing and chocolate products in the United States. *J. Agric. Food Chem.* **2006**, *54*, 4062–4068. [CrossRef]
9. Payne, M.J.; Hurst, W.J.; Stuart, D.A.; Ou, B.; Fan, E.; Ji, H.; Kou, Y. Determination of total procyanidins in selected chocolate and confectionery products using DMAC. *J. AOAC Int.* **2010**, *93*, 89–96.
10. Martínez, R.; Torres, P.; Meneses, M.; Figueroa, J.; Pérez-Álvarez, J.; Viuda-Martos, M. Chemical, technological and in vitro antioxidant properties of cocoa (*Theobroma cacao* L.) co-products. *Food Res. Int.* **2012**, *49*, 39–45. [CrossRef]
11. Oddoye, E.O.K.; Agyente-Badu, C.K.; Gyedu-Akoto, E. Cocoa and its by-products: Identification and utilization. In *Chocolate in Health and Nutrition*; Watson, R., Preedy, V., Zibadi, S., Eds.; Humana Press: Totowa, NJ, USA, 2013; Volume 7, pp. 23–37.

12. Pérez, E.; Méndez, A.; León, M.; Hernández, G.; Sívoli, L. Proximal composition and the nutritional and functional properties of cocoa by-products (pods and husks) for their use in the food industry. In *Chocolate Cocoa Byproducts Technology, Rheology, Styling, and Nutrition*; Pérez, E., Ed.; Nova Science Publishers, Inc.: New York, NY, USA, 2015; pp. 219–234.

13. Ortega, N.; Romero, M.-P.; Macia, A.; Reguant, J.; Anglès, N.; Morelló, J.; Motilva, M.-J. Comparative study of UPLC–MS/MS and HPLC–MS/MS to determine procyanidins and alkaloids in cocoa samples. *J. Food Compos. Anal.* **2010**, *23*, 298–305. [CrossRef]

14. Counet, C.; Collin, S. Effect of the number of flavanol units on the antioxidant activity of procyanidin fractions isolated from chocolate. *J. Agric. Food Chem.* **2003**, *51*, 6816–6822. [CrossRef] [PubMed]

15. Rodriguez-Ramiro, I.; Martin, M.A.; Ramos, S.; Bravo, L.; Goya, L. Comparative effects of dietary flavanols on antioxidant defences and their response to oxidant-induced stress on Caco2 cells. *Eur. J. Nutr.* **2011**, *50*, 313–322. [CrossRef] [PubMed]

16. Matias, A.A.; Rosado-Ramos, R.; Nunes, S.L.; Figueira, I.; Serra, A.T.; Bronze, M.R.; Santos, C.N.; Duarte, C.M. Protective effect of a (Poly)phenol-rich extract derived from sweet cherries culls against oxidative cell damage. *Molecules* **2016**, *21*, 406. [CrossRef] [PubMed]

17. Felice, F.; Zambito, Y.; Di Colo, G.; D'Onofrio, C.; Fausto, C.; Balbarini, A.; Di Stefano, R. Red grape skin and seeds polyphenols: Evidence of their protective effects on endothelial progenitor cells and improvement of their intestinal absorption. *Eur. J. Pharm. Biopharm.* **2012**, *80*, 176–184. [CrossRef]

18. Balestrieri, M.L.; Schiano, C.; Felice, F.; Casamassimi, A.; Balestrieri, A.; Milone, L.; Servillo, L.; Napoli, C. Effect of low doses of red wine and pure resveratrol on circulating endothelial progenitor cells. *J. Biochem.* **2008**, *143*, 179–186. [CrossRef]

19. Lin, X.L.; Liu, Y.; Liu, M.; Hu, H.; Pan, Y.; Fan, X.J.; Hu, X.M.; Zou, W.W. Inhibition of hydrogen peroxide-induced human umbilical vein endothelial cells aging by allicin depends on Sirtuin1 activation. *Med. Sci. Monit.* **2017**, *23*, 563–570. [CrossRef]

20. Felice, F.; Maragò, E.; Sebastiani, L.; Di Stefano, R. Apple juices from ancient Italian cultivars: A study on mature endothelial cells model. *Fruits* **2015**, *70*, 361–369. [CrossRef]

21. Hafizah, A.H.; Zaiton, Z.; Zulkhairi, A.; Mohd Ilham, A.; Nor Anita, M.M.; Zaleha, A.M. Piper sarmentosum as an antioxidant on oxidative stress in human umbilical vein endothelial cells induced by hydrogen peroxide. *J. Zhejiang Univ. Sci. B* **2010**, *11*, 357–365. [CrossRef]

22. Berni, R.; Romi, M.; Cantini, C.; Hausman, J.-F.; Guerriero, G.; Cai, G. Functional molecules in locally-adapted crops: The case study of tomatoes, onions, and sweet cherry fruits from Tuscany in Italy. *Front. Plant Sci.* **2019**, *9*, 1983. [CrossRef]

23. Hammerstone, J.F.; Lazarus, S.A.; Mitchell, A.E.; Rucker, R.; Schmitz, H.H. Identification of procyanidins in cocoa (*Theobroma cacao*) and chocolate using high-performance liquid chromatography/mass spectrometry. *J. Agric. Food Chem.* **1999**, *47*, 490–496. [CrossRef] [PubMed]

24. Bruna, C.; Eichholz, I.; Rohn, S.; Kroh, L.; Huyskens-Keil, S. Bioactive compounds and antioxidant activity of cocoa hulls (*Theobroma cacao* L.) from different origins. *J. Appl. Botany Food Qual.* **2009**, *83*, 9–13.

25. Kim, H.; Keeney, P. (-)-Epicatechin Content in Fermented and Unfermented Cocoa Beans. *J. Food Sci.* **2006**, *49*, 1090–1092. [CrossRef]

26. Urbanska, B.; Kowalska, J. Comparison of the total polyphenol content and antioxidant activity of chocolate obtained from roasted and unroasted cocoa beans from different regions of the World. *Antioxidants* **2019**, *8*, 283. [CrossRef] [PubMed]

27. Racine, K.C.; Wiersema, B.D.; Griffin, L.E.; Essenmacher, L.A.; Lee, A.H.; Hopfer, H.; Lambert, J.D.; Stewart, A.C.; Neilson, A.P. Flavanol polymerization is a superior predictor of alpha-glucosidase inhibitory activity compared to flavanol or total polyphenol concentrations in cocoas prepared by variations in controlled fermentation and roasting of the same raw cocoa beans. *Antioxidants* **2019**, *8*, 635. [CrossRef] [PubMed]

28. Żyżelewicz, D.; Krysiak, W.; Oracz, J.; Sosnowska, D.; Budryn, G.; Nebesny, E. The influence of the roasting process conditions on the polyphenol content in cocoa beans, nibs and chocolates. *Food Res. Int.* **2016**, *89*, 918–929. [CrossRef]

29. Beconcini, D.; Fabiano, A.; Zambito, Y.; Berni, R.; Santoni, T.; Piras, A.M.; Di Stefano, R. Chitosan-based nanoparticles containing cherry extract from *Prunus avium* L. to improve the resistance of endothelial cells to oxidative stress. *Nutrients* **2018**, *10*. [CrossRef]

30. Braca, A.; Sinisgalli, C.; De Leo, M.; Muscatello, B.; Cioni, P.L.; Milella, L.; Ostuni, A.; Giani, S.; Sanogo, R. Phytochemical profile, antioxidant and antidiabetic activities of *Adansonia digitata* L. (Baobab) from Mali, as a source of health-promoting compounds. *Molecules* **2018**, *23*. [CrossRef]
31. Benzie, I.F.; Strain, J.J. The ferric reducing ability of plasma (FRAP) as a measure of "antioxidant power": The FRAP assay. *Anal. Biochem.* **1996**, *239*, 70–76. [CrossRef]
32. Singleton, V.L.; Orthofer, R.; Lamuela-Raventos, R.M. Analysis of total phenols and other oxidation substrates and antioxidants by means of Folin–Ciocalteu reagent. *Methods Enzymol.* **1999**, *299*, 152–178.
33. Beconcini, D.; Fabiano, A.; Di Stefano, R.; Macedo, M.H.; Felice, F.; Zambito, Y.; Sarmento, B. Cherry extract from *Prunus avium* L. to improve the resistance of endothelial cells to oxidative stress: Mucoadhesive chitosan vs. poly(lactic-co-glycolic acid) nanoparticles. *Int. J. Mol. Sci.* **2019**, *20*. [CrossRef] [PubMed]
34. Jaffe, E.A.; Nachman, R.L.; Becker, C.G.; Minick, C.R. Culture of human endothelial cells derived from umbilical veins. Identification by morphologic and immunologic criteria. *J. Clin. Investig.* **1973**, *52*, 2745–2756. [CrossRef] [PubMed]
35. Fabiano, A.; Mattii, L.; Braca, A.; Felice, F.; Di Stefano, R.; Zambito, Y. Nanoparticles based on quaternary ammonium-chitosan conjugate: A vehicle for oral administration of antioxidants contained in red grapes. *J. Drug Deliv. Technol.* **2016**, *32*, 291–297. [CrossRef]
36. Legen, I.; Salobir, J.; Kerc, J. Comparison of different intestinal epithelia as models for absorption enhancement studies. *Int. J. Pharm.* **2005**, *291*, 183–188. [CrossRef] [PubMed]
37. Cádiz-Gurrea, M.L.; Lozano-Sanchez, J.; Contreras-Gámez, M.; Legeai-Mallet, L.; Fernández-Arroyo, S.; Segura-Carretero, A. Isolation, comprehensive characterization and antioxidant activities of Theobroma cacao extract. *J. Funct. Foods* **2014**, *10*, 485–498. [CrossRef]
38. Tomas-Barberan, F.A.; Cienfuegos-Jovellanos, E.; Marin, A.; Muguerza, B.; Gil-Izquierdo, A.; Cerda, B.; Zafrilla, P.; Morillas, J.; Mulero, J.; Ibarra, A.; et al. A new process to develop a cocoa powder with higher flavonoid monomer content and enhanced bioavailability in healthy humans. *J. Agric. Food Chem.* **2007**, *55*, 3926–3935. [CrossRef] [PubMed]
39. Rue, E.A.; Rush, M.D.; van Breemen, R.B. Procyanidins: A comprehensive review encompassing structure elucidation via mass spectrometry. *Phytochem. Rev.* **2018**, *17*, 1–16. [CrossRef]
40. Ortega, N.; Romero, M.P.; Macia, A.; Reguant, J.; Angles, N.; Morello, J.R.; Motilva, M.J. Obtention and characterization of phenolic extracts from different cocoa sources. *J. Agric. Food Chem.* **2008**, *56*, 9621–9627. [CrossRef]
41. Zambito, Y.; Zaino, C.; Uccello-Barretta, G.; Balzano, F.; Di Colo, G. Improved synthesis of quaternary ammonium-chitosan conjugates (N^+-Ch) for enhanced intestinal drug permeation. *Eur. J. Pharm. Sci.* **2008**, *33*, 343–350. [CrossRef]
42. Kumar, V.; Khan, A.A.; Tripathi, A.; Dixit, P.K.; Bajaj, U.K. Role of oxidative stress in various diseases: Relevance of dietary antioxidants. *J. Pharm. Exp. Ther.* **2015**, *4*, 126–132.
43. Usenik, V.; Fabčič, J.; Stampar, F. Sugars, organic acids, phenolic composition and antioxidant activity of sweet cherry (*Prunus avium* L.). *Food Chem.* **2008**, *107*, 185–192. [CrossRef]
44. Rebollo-Hernanz, M.; Zhang, Q.; Aguilera, Y.; Martín-Cabrejas, M.A.; Gonzalez de Mejia, E. Relationship of the phytochemicals from coffee and cocoa by-products with their potential to modulate biomarkers of metabolic syndrome in vitro. *Antioxidants* **2019**, *8*, 279. [CrossRef] [PubMed]
45. Jung, K.A.; Kwak, M.K. The Nrf2 system as a potential target for the development of indirect antioxidants. *Molecules* **2010**, *15*, 7266–7291. [CrossRef] [PubMed]
46. Galvano, F.; Fauci, L.; Lazzarino, G.; Fogliano, V.; Ritieni, A.; Ciappellano, S.; Battistini, N.C.; Tavazzi, B.; Galvano, G. Cyanidins: Metabolism and biological properties. *J. Nutr. Biochem.* **2004**, *15*, 2–11. [CrossRef]
47. Zhao, C.-N.; Meng, X.; Li, Y.; Li, S.; Liu, Q.; Tang, G.-Y.; Li, H.-B. Fruits for Prevention and Treatment of Cardiovascular Diseases. *Nutrients* **2017**, *9*, 598. [CrossRef]

 *antioxidants*

Article

(–)-Epicatechin Reduces the Blood Pressure of Young Borderline Hypertensive Rats During the Post-Treatment Period

Michal Kluknavsky [1], Peter Balis [1], Martin Skratek [2], Jan Manka [2] and Iveta Bernatova [1,*]

[1] Slovak Academy of Sciences, Centre of Experimental Medicine, Institute of Normal and Pathological Physiology, 813 71 Bratislava, Slovakia; michal.kluknavsky@savba.sk (M.K.); peter.balis@savba.sk (P.B.)

[2] Slovak Academy of Sciences, Institute of Measurement Science, 841 04 Bratislava, Slovakia; martin.skratek@savba.sk (M.S.); jan.manka@savba.sk (J.M.)

* Correspondence: iveta.bernatova@savba.sk

Received: 17 December 2019; Accepted: 19 January 2020; Published: 23 January 2020

Abstract: This study investigated the effects of (–)-epicatechin (Epi) in young male borderline hypertensive rats (BHR) during two weeks of treatment (Epi group, 100 mg/kg/day p.o.) and two weeks post treatment (PE group). Epi reduced blood pressure (BP), which persisted for two weeks post treatment. This was associated with delayed reduction of anxiety-like behaviour. Epi significantly increased nitric oxide synthase (NOS) activities in the aorta and left heart ventricle (LHV) vs. the age-matched controls without affecting the brainstem and frontal neocortex. Furthermore, Epi significantly reduced the superoxide production in the aorta and relative content of iron-containing compounds in blood. Two weeks post treatment, the NOS activities and superoxide productions in the heart and aorta did not differ from the age-matched controls. The gene expressions of the NOSs (*nNOS, iNOS, eNOS*), nuclear factor erythroid 2-related factor 2 (*Nrf2*), and peroxisome proliferator-activated receptor-γ (*PPAR-γ*) remained unaltered in the aorta and LHV of the Epi and PE groups. In conclusion, while Epi-induced a decrease of the rats' BP persisted for two weeks post treatment, continuous Epi treatments seem to be necessary for maintaining elevated NO production as well as redox balance in the heart and aorta without changes in the *NOSs, Nrf2*, and *PPAR-γ* gene expressions.

Keywords: (–)-epicatechin; borderline hypertensive rats; nitric oxide; redox balance; iron; *Nrf2*; *PPAR-γ*; open field

1. Introduction

Borderline hypertension, i.e., prehypertension, was first defined in the seventh report of the Joint National Committee on Prevention, Detection, Evaluation, and Treatment of High Blood Pressure, as being characterized by systolic blood pressure (BP) between 120 and 139 mm Hg or diastolic BP between 80 and 89 mm Hg [1]. Data from the 1999–2008 that were published by the National Heart, Lung, and Blood Institute showed the prevalence of prehypertension in the young US population (18–29 years old) ~ 26% and ~33% among those aged 40–49 years [2]. High prevalence of prehypertension has been found in school-aged and adolescent populations [3]. In addition, several studies have confirmed the elevated risk of cardiovascular diseases in prehypertensive people [4,5]. These findings have led to the formulation of new guidelines for the prevention, evaluation, and management of people with systolic BP above 120–139 mm Hg.

There are several terms for systolic BP that are in the range of 120–139 mm Hg. The European Society of Cardiology and European Society of Hypertension defined systolic BP of 120–129 mm Hg as "normal" and 130–139 mm Hg as "high normal" [6]. The American College of Cardiology and American Heart Association defined systolic BP of 120–129 mm Hg as "elevated" and 130–139 mm

Hg as "stage 1 hypertension" [7]. Both of the guidelines currently recommend non-pharmacological interventions for each stage of elevated systolic BP and pharmacological treatment for high normal/stage 1 hypertension (i.e., systolic BP in the range of 130–139 mm Hg) if other risk factors are present, despite the differing terminologies [6,7]. There is a considerably lower number of experimental studies focused on the mechanism of preventing high BP during the prehypertensive phase than studies focused on the treatment of fully developed hypertension, despite the high prevalence of prehypertension among humans.

One of the experimental models of essential prehypertension consists of young spontaneously hypertensive rats (SHR). Several studies have showed that early pharmacological treatment in young SHR during the prehypertensive period (4–8 weeks of age) has a prolonged BP-lowering effect and it maintains cardiac protection in adulthood [8,9]. However, young SHR are not a suitable model for representing essential prehypertension in adulthood due to the early onset of fully developed hypertension (i.e., systolic BP over 140 mm Hg) before reaching adulthood (~12 weeks of age) [10]. A more appropriate genetic model of essential prehypertension consists of borderline hypertensive rats (BHR), which were obtained by the mating of female SHR and male normotensive Wistar-Kyoto (WKY), resulting in a systolic BP of approximately 140 mm Hg in the adulthood stage [11,12].

Regarding the mechanisms of (pre)hypertension development, several human studies noticed increased markers of oxidative stress and decreased antioxidant capacity in the plasma of young as well as middle-aged prehypertensive adults [13,14]. Oxidative damage was also observed in adult 22-week-old BHR [11], which suggested that the altered redox state might contribute to the development of (pre)hypertension. Iron, which is an essential nutrient for all cells, in addition to its other functions, contributes to the production of reactive oxygen species (ROS), modulates redox balance [15], and affects nitric oxide bioavailability and, thus, vascular function [16]. However, the role of iron and/or iron-containing compounds in the development of arterial hypertension is not fully understood. For example, studies of Astma et al. [17] and Zhu et al. [18] showed that the haemoglobin levels were positively associated with BP in non-hypertensive people. In addition, a high incidence of systolic BP in prehypertensive range prevailed in young adults, in them haemoglobin level was positively correlating with systolic BP [19]. In contrast, in patients with hereditary haemochromatosis with profound iron overload endothelial function was impaired without changes in BP [20]. Endothelial dysfunction, which results from reduced nitric oxide (NO) bioavailability, is another factor that is involved in the development of (pre)hypertension [21]. In addition, alterations in the renin-angiotensin system and in neurogenic regulation also contribute to the development of (pre)hypertension [22].

Several lifestyle-related changes are recommended to reduce elevated BP, such as weight reduction, exercise, and a healthy diet. High attention is paid to the consumption of flavan-3-ols (catechins), as several clinical and epidemiological studies have found that a reduction in the systolic BP is associated with an increased consumption of cocoa products [23], the rich source of Epi. The recent studies have confirmed that Epi is the subject of major in vivo metabolisation and Epi and/or its metabolites are absorbed well from the gastrointestinal tract [24,25]. Furthermore, there are other studies that found Epi and/or its metabolites in plasma in concentrations that are sufficient to produce biological effects [23]. A decrease in the BP, dependent on the Epi content, was found in a meta-analysis that was performed by Ellinger et al. [26]. The BP-lowering effect of Epi was also observed in various experimental studies while using with N(ω)-nitro-L-arginine methyl ester (L-NAME)-induced [27], fructose-induced [28], and deoxycorticosterone acetate (DOCA)–salt-induced hypertension [29]. Epi was also found to induce a reduction in the systolic BP of both young and adult SHR [30,31]. Elevated NOS activity and reduced oxidative stress seem to be the main factors that lead to a BP decrease after long-term Epi treatment.

However, little is known about the molecular mechanisms that are associated with the elevated NO production and reduced release of ROS during Epi treatments. Nuclear factor erythroid 2-related factor 2 (Nrf2) is a transcription factor that plays a significant role in hypertension development [32,33]. Nrf2 is involved in cellular responses to oxidative stress; it modifies the ROS level by modulating the antioxidant and detoxification outputs as well as it modifies inflammation [33–35]. Peroxisome proliferator-activated receptor gamma (PPAR-γ) is transcription factor that modulates several signalling pathways, such as phosphoinositide 3-kinase/protein kinase B/nitric oxide synthase, the renin-angiotensin system, angiotensin-receptor 1/nicotinamide adenine dinucleotide phosphate oxidase pathway, as well as redox homeostasis [36]. Nrf2 and PPAR-γ are both present in the heart and arteries [37,38] and it might be involved in the Epi-induced BP-lowering effect.

Therefore, this study investigated the effects of (−)-epicatechin on BP, locomotor activity, anxiety-like behaviour, nitric oxide synthase (NOS) activity, and superoxide ($O_2\bullet^-$) production in the aorta, left heart ventricle (LHV), brainstem, and frontal neocortex, as well as on the relative content of iron-containing compounds in the blood of young borderline BHR. In addition, the gene expressions of NO synthase isoforms (*nNOS, eNOS, iNOS*) and involvement of mechanisms that are mediated by transcription factors Nrf2 and PPAR-γ were investigated in the LHV and aorta. Furthermore, we tested the hypothesis that Epi treatment can result in a long-term reduction of BP, behavioural alterations, and alterations in the abovementioned genes after the cessation of Epi administration.

2. Materials and Methods

2.1. Animals and Treatment

Five-week-old BHR males, i.e., the offspring of SHR dams and WKY sires, were used in this study. The rats ($n = 28$) were divided into four groups: the seven-week control (C7, $n = 7$), Epi-treated (Epi, $n = 7$), nine-week control (C9, $n = 7$), and post-Epi (PE, $n = 7$) groups. The rats in the Epi and PE groups were treated with Epi (Sigma-Aldrich, Bratislava, Slovakia, Cat. no. E1753) added to drinking water for two weeks. Epi was prepared fresh every day before administration to rats by suspension of Epi in tap water at ~85 °C for 5 min. while using vortex. The calculated volume of this Epi, based on the body weight of rats and drinking volume, was added to the drinking bottles of the rats to reach a final dose of approximately 100 mg/kg/day after drinking all of the liquid during the 24 h period. The rats in both the control groups drank tap water. Rats in the PE group drank tap water for two weeks after the cessation of the Epi administration. The dose of Epi was selected on the basis of our previous study in age-matched SHR, in which the same dose and way of Epi administration elevated antioxidant capacity of plasma and improved aortic NO bioavailability when determined as the elevated NO-dependent component of acetylcholine-induced relaxation [30].

The body weights (BW) of all the rats were measured on the same day as the BP measurements. At the end of the experiment, all of the rats were exposed to brief CO_2 anaesthesia and subsequently killed by decapitation. After decapitation, the relative weight of the left heart ventricle (LHV/BW) was calculated. Increasing the LHV/BW ratio served as a marker of left-heart hypertrophy. We also measured the relative weight of the left and right kidneys (LK + RK/BW). The Department of Animal Wellness, State Veterinary and Food Administration of the Slovak Republic approved all of the procedures in accordance with the EU Directive 2010/63/EU, decision No. Ro-2561/12-221.

2.2. Open-Field Test

Locomotor activity and anxiety-like behaviour were measured while using the open-field test (OF) between 07:30 a.m. and 10:00 a.m. with a video tracking system called Any-maze (Stoelting, Ireland). At the beginning of the experiment, all of the rats were tested at the age of five weeks (Basal, $n = 28$). After the basal behavioural testing, the rats were assigned to the control (C7, C9) or Epi-treated groups (Epi, PE). The behavioural tests were conducted on day 13 for the C7 and Epi groups and day 27 for the C9 and PE groups. Kluknavsky et al. have previously described the testing conditions in detail [30]. The total distance travelled and total time of immobility in the OF were determined as the parameters of locomotor activity. As markers of anxiety-like behaviour, total distance and relative distance travelled in the central zone (calculated as the percentage of central zone distance with respect to the total distance travelled) were determined.

2.3. Systolic Blood Pressure

The systolic BP was measured in preconditioned, conscious rats while using non-invasive tail-cuff plethysmography between 08:00 a.m. and 11:00 a.m., as previously described by Puzserova et al. [39]. Each value was calculated as the average of five measurements. The BP values were measured at the beginning of the experiment (Basal), and then on the days 3, 7, 14, 17, 21, and 28 of the experiment.

2.4. Determination of the Relative Iron Content in Blood

A Quantum Design MPMS-XL 7AC (SQUID) magnetometer with a reciprocating sample operation option (with differential sensitivity of 10^{-11} Am^2 up to 0.25 T and 10^{-10} Am^2 up to 7 T) was used to compare the relative content of iron and iron-containing compounds in the blood samples. The advantage of this method is its ability to determine all of the iron forms in a small sample of biological material, with high sensitivity [40–42].

After decapitation, trunk blood was collected in Eppendorf test tubes and then stored at −80 °C. Subsequently, the blood was defrosted and homogenised using an ultrasonic bath for 60 s (50 kHz, 30 W). After homogenisation 10 µL of the blood was placed on a 16 cm × 6 mm strip of standard white office paper (80 g/m^2), vacuum-dried, and inserted into a plastic measuring tube [42]. Thus, we obtained a dry sample of the blood in a homogeneous ambient without the need for a gelatine capsule, which is usually the source of an additional magnetic signal. The magnetic characteristics of the samples were measured in the form of isothermal hysteresis curves [40] at a temperature of 2 K (−271.15 °C) and at a magnetic field up to 7 T when the saturation magnetisation (M_s) was reached. M_s is the parameter for determining the concentration of magnetic particles in the biological samples, of which iron is a dominant component under the conditions used in this study. Changes in the magnetic forms of iron, which result from the various sizes of the iron-containing compounds, can be characterised by alterations in the remanent magnetisation (M_r) and coercivity (H_c) [43]. The magnetic properties of the blood were determined for the C7 and Epi groups ($n = 5$ per group).

2.5. Superoxide Production

The production of superoxide was measured in the LHV and thoracic aorta (15–20 mg) samples with lucigenin (50 µmol/L)-enhanced chemiluminescence while using a TriCarb 2910TR liquid scintillation analyser (Perkin Elmer, Waltham, MA, USA), as described previously by Slezak et al. [44]. The results are expressed in the form of cpm/mg of tissue.

2.6. Nitric Oxide Synthase Activity

The total NOS activity was measured in the 20% tissue homogenates of the LHV, aorta, brainstem, and frontal neocortex by determining the [^{3}H]-L-citrulline formation from [^{3}H]-L-arginine (ARC, St. Louis, MO, USA), as described previously [39], and expressed in terms of pmol/min./mg of tissue protein. The protein concentration was determined while using the Lowry method.

2.7. Gene Expression

The expression levels of *Nrf2*, *PPAR-γ*, neuronal NOS (*nNOS*, isoform I), inducible NOS (*iNOS*, isoform II), endothelial NOS (*eNOS*, isoform III), and *ß-actin* (housekeeping gene) were determined while using a real-time quantitative polymerase chain reaction (RT-qPCR). The total RNA of the thoracic aorta, LHV, brainstem and frontal neocortex were isolated using the TRIsure reagent (Bioline, United Kingdom), according to the manufacturer's protocols. The amount of total isolated RNA was spectrophotometrically quantified at 260/280 nm while using a NanoDrop spectrophotometer (Thermo Scientific, Waltham, MA, USA). Reverse transcription was performed using 1 µg of the total RNA of each sample and 20 µL of the reaction medium using a SensiFAST™ cDNA Synthesis Kit (Bioline, London, United Kingdom), in accordance with the manufacturer's protocols (Eppendorf Mastercycler, Hamburg, Germany). SensiFAST mix (SensiFAST SYBR No-ROX kit, Bioline, London, UK) was used for gene amplification. The preparation of the PCR mixture, thermal cycling conditions, used apparatus, and detection software were described previously [30]. Gene-specific primers were designed while using the PubMed database (Gene) and program (Primer-BLAST). The primer pairs used to amplify the nNOS (GenBank accession No. NM_052799.1), iNOS (GenBank accession No. NM_012611.3), eNOS (GenBank accession No. NM_021838.2), PPAR-γ (GenBank accession No. NM_013124.3) and Nrf2 (GenBank accession No. NM_031789.2) genes, as well as β-actin (GenBank accession No. NM_031144.3, a housekeeping gene) are listed in Table 1. The samples were measured using the Bio-Rad CFX Manager software 2.0 (Hercules, CA, USA). The gene expressions were considered as the ratio of the given gene's expression to β-actin expression.

All of the chemicals used in this study were purchased from Sigma-Aldrich (Bratislava, Slovakia) and Merck Chemicals (Bratislava, Slovakia), unless stated otherwise.

Table 1. Primer pairs used to amplify selected genes.

Gene	Forward (Sense) Primer	Reverse (Antisense) Primer	Temp
eNOS	CGGCGCAAAAGGAAGGAATC	CCAGCCCAAACACACAGAAC	60 °C
iNOS	TGGAGGTGCTGG AAGAGTT	GGAGGAGCTGATGGAGTAGT	57 °C
nNOS	CGCTACGCGGGCTACAAGCA	GCACGTCGAAGCGGCCTCTT	60 °C
Nrf2	AGGTTGCCCACATTCCCAAA	TATCCAGGGCAAGCGACTCA	60 °C
PPAR-γ	TCCCGTTCACAAGAGCTGAC	GCTCTACTTTGATCGCACTTTGG	60 °C
β-actin	AATCGTGCGTGACATCAAAG	ATGCCACAGGATTCCATACC	57 °C

2.8. Statistical Analysis

The BP results were separately analysed for the Epi-treatment period (n = 14 in each group) and post-treatment period (n = 7 in each group) while using a two-way ANOVA (treatment x day of experiment). The saturation magnetisation of the blood was analysed using the Student's t-test. All of the other data were analysed with a one-way ANOVA. All of the ANOVA analyses were followed by Bonferroni post-hoc tests. The correlations between the gene expressions were determined using Pearson's correlation coefficient (r). The values were found to significantly differ when $p < 0.05$. The data were presented in the form of mean ± standard error of the mean (SEM). GraphPad Prism 5.0 (GraphPad Software, Inc., San Diego, CA, USA) was used for the statistical analyses.

3. Results

There were age-dependent changes in the BW and relative kidney mass (LK + RK/BW), and Epi had no effect on these parameters when compared to the age-matched controls (Table 2). No alterations were found in the relative LHV mass (LHV/BW) among the groups (Table 2). The saturation magnetisation of the Epi-treated rats was significantly reduced by approximately 54% vs. the age-matched control group. No significant changes were found in the remanent magnetisation and coercivity (Table 2).

Table 2. Basic biometric parameters and magnetic properties of blood.

Group	BW (g) $n = 7$	LHV/BW (mg/g) $n = 7$	(LK + RK)/BW (mg/g) $n = 7$	M_s (10^{-3} Am^2/kg) $n = 5$	M_r (10^{-2} Am^2/kg) $n = 5$	H_c (A/m) $n = 5$
C7	178 ± 5.9	1.98 ± 0.10	9.33 ± 0.18	210 ± 13.85	0.19 ± 0.07	1569 ± 727
Epi	177 ± 2.1	1.89 ± 0.09	9.40 ± 0.26	98 ± 26.26 [x]	0.08 ± 0.01	1260 ± 244
C9	249 ± 4.6 [x]	1.84 ± 0.04	8.07 ± 0.09 [x]	N.D.	N.D.	N.D.
PE	244 ± 4.0 [+]	1.95 ± 0.05	8.11 ± 0.05 [+]	N.D.	N.D.	N.D.

Abbreviations: C7: seven-week-old control group, C9: nine-week-old control group, Epi: (–)-epicatechin group, PE: (–)-epicatechin post-treatment group, BW: body weight, LHV: left heart ventricle, LK: left kidney, RK: right kidney, M_s: saturation magnetisation of the blood samples, M_r: remanent magnetisation of blood samples, H_c: coercivity of blood samples, N.D.: not determined. The values represent the mean ± SEM. [x] $p < 0.001$ vs. C7, [+] $p < 0.001$ Epi.

There were no significant differences in the basal BP of rats that were assigned to the control groups (132 ± 2 mm Hg, $n = 14$) and Epi-treated groups (137 ± 2 mm Hg, $n = 14$) at the beginning of the study (Figure 1). The ANOVA analysis confirmed significant differences in BP between the groups during the treatment ($F_{1,100} = 19.03$, $p < 0.0001$, main effect of treatment), the significant effect of the day of treatment ($F_{3,100} = 4.08$, $p < 0.01$, main effect of day), as well as the significant effect of the interaction of treatment and day of experiment ($F_{3,100} = 5.80$, $p < 0.002$) during the treatment period. Epi significantly reduced the BP during the treatment when compared to the control group. During the post-treatment period (i.e., days 17, 21, and 28), the ANOVA showed significant difference in BP between the control and PE groups ($F_{1,35} = 10.65$, $p < 0.003$, main effect of post-treatment) and the significant effect of the day of experiment ($F_{2,35} = 6.95$, $p < 0.003$, main effect of day).

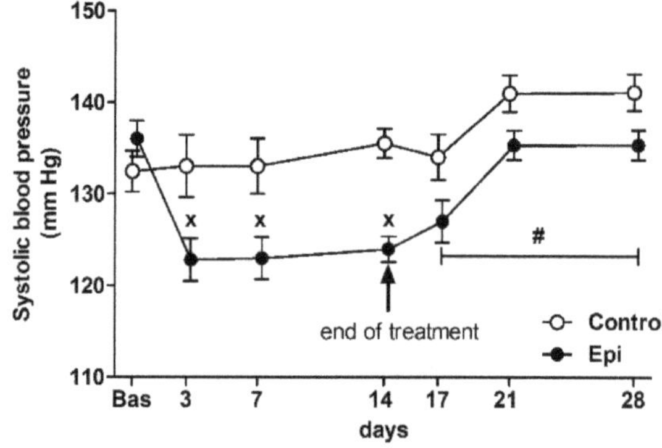

Figure 1. The effect of (–)-epicatechin on the systolic blood pressure of borderline hypertensive rats. The values represent the mean ± SEM. [x] $p < 0.05$ vs. control group on the respective day of experiment, [#] $p < 0.03$ vs. control group (ANOVA main effect of group during post-treatment period i.e., on days 17, 21 and 28, see Results). Abbreviations. Bas: baseline value, Epi: (–)-epicatechin group.

The ANOVA analysis confirmed significant differences in OF behaviour during testing for all of the distance travelled ($F_{4,59} = 7.76$, $p < 0.001$), immobility ($F_{4,59} = 12.5$, $p < 0.001$), distance travelled in the central zone ($F_{4,59} = 3.83$, $p < 0.008$), and relative distance travelled in the central zone ($F_{4,59} = 2.9$, $p < 0.03$). Repeated testing of BHR in OF led to the habituation of locomotor activity and immobility (Figure 2a,b) when compared to the respective basal values. There were no differences in the individual values determined for the Epi group as compared to the age-matched controls (C7). However, for the PE group, a lack of habituation (vs. the basal value) was found in distance travelled in the central zone, in contrast to the age-matched C9 control group (Figure 2c). This resulted in the significant increase in relative distance travelled in the central zone for the PE group as compared to the age-matched (C9) control group (Figure 2d).

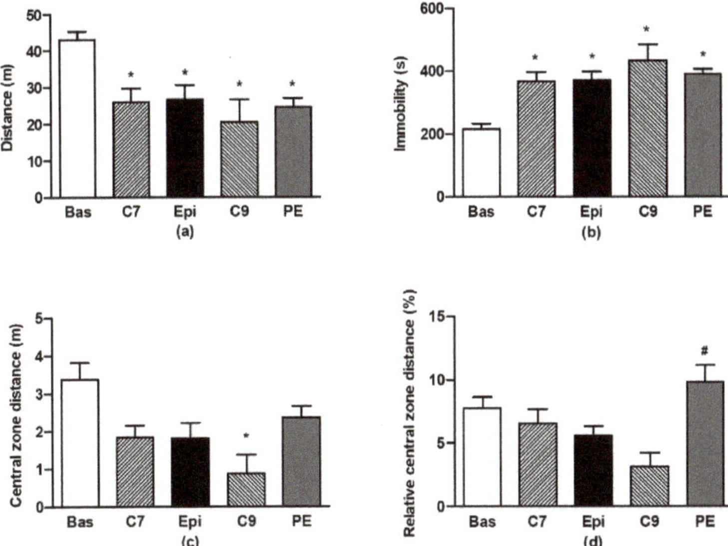

Figure 2. The effect of (–)-epicatechin on the behaviours of borderline hypertensive rats. Total distance travelled (**a**) and total immobility (**b**) in the open field; total distance (**c**) and relative distance (**d**) travelled in the central zone of the open field. The values represent the mean ± SEM. * $p < 0.05$ vs. the basal group value, # $p < 0.05$ vs. the C9 group. Abbreviations. C7: seven-week-old control group, C9: nine-week-old control group, Epi: (–)-epicatechin group, PE: (–)-epicatechin post-treatment group.

In the LHV, Epi did not change $O_2\bullet^-$ production as compared to the age-matched control groups (Figure 3a). In the aorta, a significant ($p < 0.001$) reduction of $O_2\bullet^-$ production was detected in the Epi group when compared to the C7 group, and a similar tendency was seen in the post-treatment period (Figure 3b).

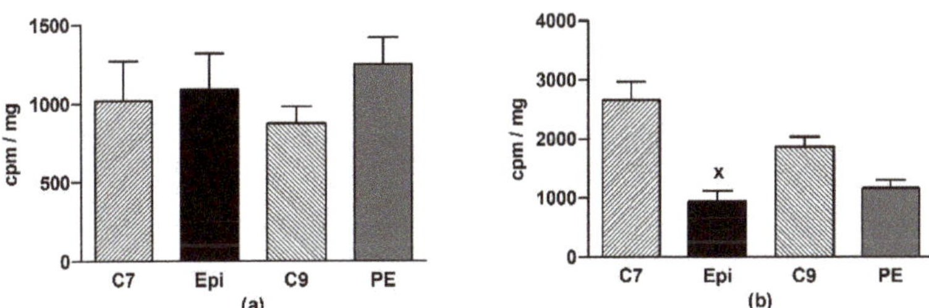

Figure 3. The effect of (–)-epicatechin on the superoxide production in the left heart ventricle (**a**) and aorta (**b**) of borderline hypertensive rats. The values represent the mean ± SEM. ˣ $p < 0.001$ vs. C7 group, $n = 6$–7 per group. Abbreviations. C7: seven-week-old control group, C9: nine-week-old control group, Epi: (–)-epicatechin group, PE: (–)-epicatechin post-treatment group.

The Epi-treatment significantly ($p < 0.05$) increased the total NOS activity in the LHV, returning back to the control levels after cessation of the treatment (Figure 4a). Epi failed to alter the gene expressions of *eNOS*, *nNOS*, *iNOS*, as well as *Nrf2* and *PPAR-γ* in the LHV (Figure 4b–f).

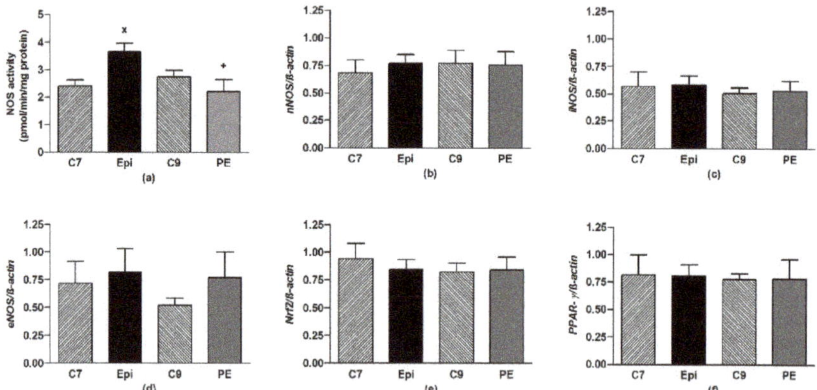

Figure 4. The effect of (–)-epicatechin on the nitric oxide synthase (NOS) activity (**a**) and gene expression of neuronal *nNOS* (**b**), inducible *iNOS* (**c**), endothelial *eNOS* (**d**), Nrf2 (**e**) and *PPAR-γ* (**f**) in the left heart ventricle of borderline hypertensive rats. The values represent the mean ± SEM, n = 6–7 per group. $^x p < 0.05$ vs. C7 group, $^+ p < 0.05$ vs. Epi group. Abbreviations. C7: seven-week-old control group, C9: nine-week-old control group, Epi: (–)-epicatechin group, PE: (–)-epicatechin post-treatment group, *Nrf2*: nuclear factor erythroid 2-related factor 2, *PPAR-γ*: peroxisome proliferator-activated receptor gamma.

Epi significantly ($p < 0.04$) increased the total NOS activity in the aorta as compared to C7 (Figure 5a), despite the gene expressions of the individual NOS isoforms remaining unchanged (Figure 5b–d). During the post-treatment period, the NOS activity and gene expressions were similar to those in the age-matched C9 control group.

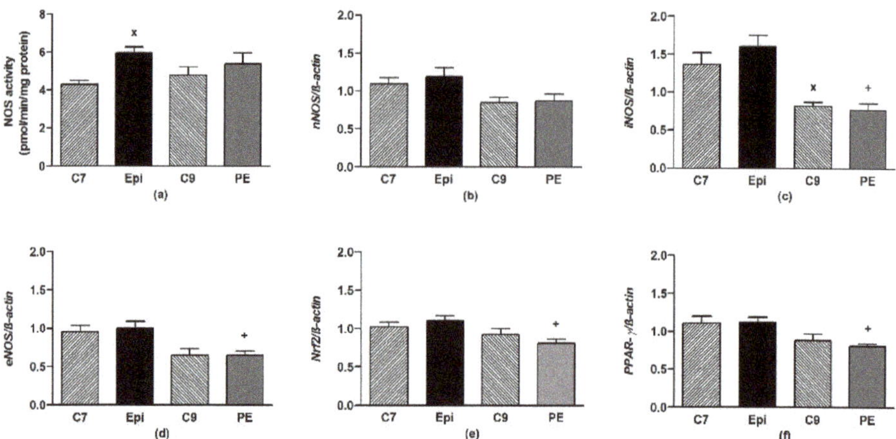

Figure 5. The effect of (–)-epicatechin on the nitric oxide synthase (NOS) activity (**a**) and gene expression of neuronal nNOS (**b**), inducible iNOS (**c**), endothelial eNOS (**d**), Nrf2 (**e**) and PPAR-γ (**f**) in the aorta of borderline hypertensive rats. The values represent the mean ± SEM; n = 6–7 per group. $^x p < 0.05$ vs. C7 group, $^+ p < 0.05$ vs. Epi group. Abbreviations: C7: seven-week-old control group, C9: nine-week-old control group, Epi: (–)-epicatechin group, PE: (–)-epicatechin post-treatment group, Nrf2: nuclear factor erythroid 2-related factor 2, PPAR-γ: peroxisome proliferator-activated receptor gamma.

There were significant correlations between the gene expressions of *Nrf2* and *PPAR-γ*, *eNOS*, *iNOS*, and *nNOS*, as shown in Table 3.

In the frontal neocortex and brainstem, Epi had no effect on the NOS activity or expression of the individual NOS isoforms as compared to the respective age-matched control group (Figures 6 and 7).

Table 3. Correlations between the gene expressions in the aorta and left heart ventricle.

	Nrf2					
	Aorta			Left Heart Ventricle		
	r	*p <*	*n*	*r*	*p <*	*n*
PPAR-γ	0.896	0.0001	26	0.714	0.0001	24
eNOS	0.772	0.0001	28	0.719	0.0001	24
nNOS	0.694	0.0001	27	0.570	0.004	24
iNOS	0.689	0.0001	27	0.508	0.01	25

The Pearson's correlation coefficients (*r*) for the gene expressions of the nuclear factor erythroid 2-related factor 2 (*Nrf2*), endothelial (*eNOS*), neuronal (*nNOS*), inducible (*iNOS*), nitric oxide synthase and peroxisome proliferator-activated receptor-γ (*PPAR-γ*), respectively.

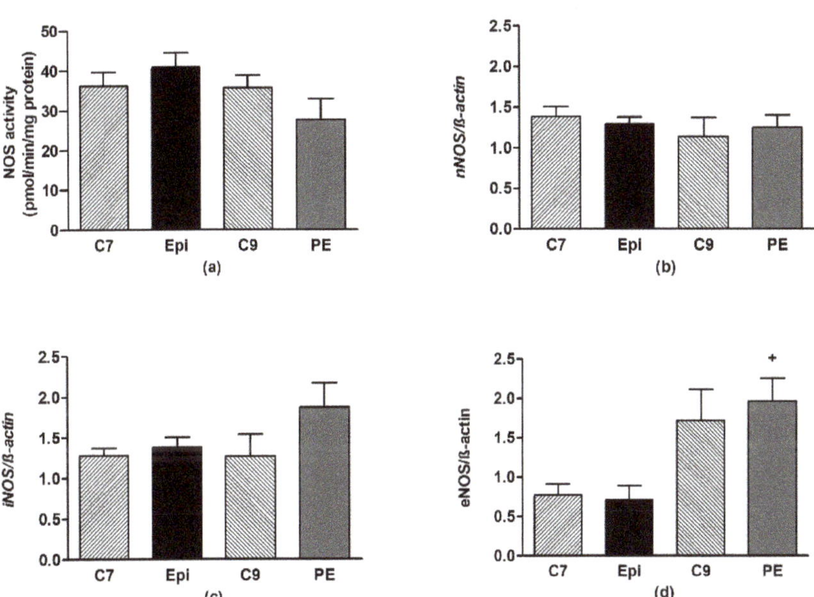

Figure 6. The effect of (–)-epicatechin on the nitric oxide synthase (NOS) activity (**a**) and gene expression of neuronal nNOS (**b**), inducible iNOS (**c**) and endothelial eNOS (**d**) in the frontal neocortex of borderline hypertensive rats. The values represent the mean ± SEM. $^{+}$ $p < 0.05$ vs. Epi group. Abbreviations. C7: seven-week-old control group, C9: nine-week-old control group, Epi: (–)-epicatechin treatment group, PE: (–)-epicatechin post-treatment group.

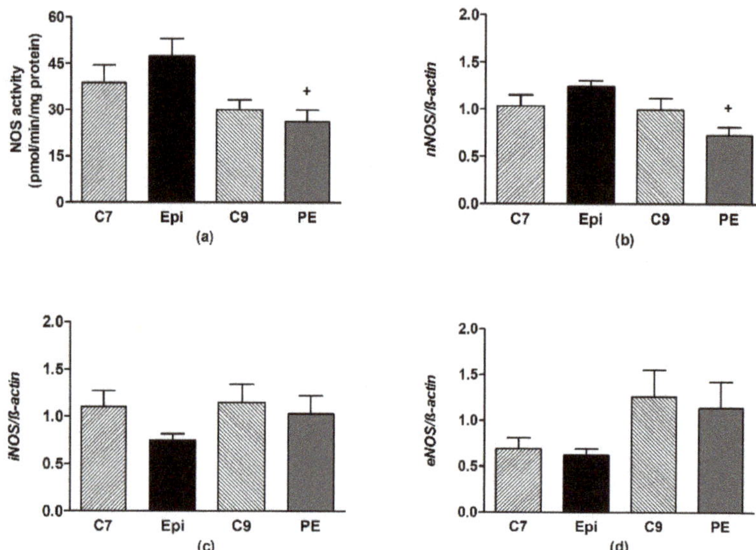

Figure 7. The effect of (–)-epicatechin on the nitric oxide synthase (NOS) activity (**a**) and gene expression of neuronal *nNOS* (**b**), inducible *iNOS* (**c**) and endothelial *eNOS* (**d**) in the brainstem of borderline hypertensive rats. The values represent the mean ± SEM. $^+$ $p < 0.05$ vs. Epi group. *Abbreviations.* C7: seven-week-old control group, C9: nine-week-old control group, Epi: (–)-epicatechin group, PE: (–)-epicatechin post-treatment group.

4. Discussion

The effects of an increased intake of cocoa products or Epi and Epi-containing foods are under investigation, not only in the area of BP regulation, but also in relation to the various neuropsychological and mood disorders in humans [23]. There are currently no studies investigating the modulation of locomotor activity and anxiety-like behaviour resulting from purified Epi or Epi-containing foods in BHR rats. BHR, which is an experimental model of human essential prehypertension, might inherit a predisposition to locomotor hyperactivity from the SHR parent [45]. However, the results of this study showed that Epi failed to significantly affect the OF behaviour of young BHR after two-week treatment in contrast to the reduced locomotor activity and relative distance travelled in the central zone previously found in the age-matched Epi-treated SHR [30]. However, the low habituation level for distance travelled in the central zone and increased relative distance travelled in the central zone at two weeks post treatment suggest a delayed anxiolytic effect of Epi in young BHR. In a study with normotensive Wistar rats, a single-dose of cocoa had an anxiolytic effect with unchanged locomotor activity, whereas a two-week treatment period failed to affect the behaviour of the rats [46]. In a study with Wistar-Unilever rats, the administration of cocoa extract for two weeks had antidepressant effects [47]. The results currently available in the literature suggest that Epi has ambiguous effects on behaviour, which probably depends on the animal model used. However, the literature has also indicated the penetration of Epi and its metabolites into the brain [48]. Further studies are required for clarifying the mechanisms of Epi's action on the spontaneous behaviour of rats and its possible association with blood pressure.

On the other hand, the BP-lowering effects of Epi are well known. In this study, the five-week BHR had a systolic BP of ~134 mm Hg, which slowly increased to ~141 mm Hg through the course of this experiment (in untreated rats). These values are similar to those that were observed in previous studies in adult BHR [11,12,49]. Regarding the development of BP in young BHR males, our results suggest the absence of a critical BP developmental window associated with the rising BP at the age of 5–7 weeks, in contrast to that in young SHR [50].

The depressor effect of Epi was observed during the entire period of treatment, followed by a gradual increase in the BP after cessation of treatment. This study is, to our knowledge, the first one that aimed at investigating the long-term effects of Epi in BHR after the cessation of the Epi treatment. The long-term influence of Epi on BP was studied on the basis of previous studies that showed long-term beneficial effects of antihypertension treatment in young prehypertensive SHR [8,9]. We found that the lower values of BP persisted two weeks after the cessation of the Epi treatment (statistically determined as the main effect), which might have been a result of the prevention of BP-induced vascular and/or organ changes that cause postponed BP increase.

Several human studies have examined the effects of flavanol-rich cocoa products associated with high content of Epi in prehypertensive individuals, despite the lack of animal studies. Petyaev et al. described the depressoric effects of cocoa products in prehypertensive subjects after two weeks of consuming 30 g dark chocolate that contained 85% cocoa [51]. Similar findings were reported for prehypertensive subjects after 15 days of consuming 30 g dark chocolate that contained 70% cocoa [52]. On the other hand, Ried et al. observed unaltered systolic BP in prehypertensive subjects after eight weeks of consuming 50 g dark chocolate that contained 70% cocoa [53]. The variability of the results that were obtained from those studies can be associated to the differing ages of the subjects, lengths of the treatment and, particularly, the variable flavanol content used in the individual studies.

Oxidative stress, which iron plays an important role in causing, can also result in the development of elevated BP [34,54,55]. In our study, Epi significantly reduced the level of iron-containing compounds in blood; reduced saturation magnetisation was observed while using SQUID magnetometry. However, the size of the iron particles and/or their chemical moiety seems to be unchanged, as coercivity and remanent magnetisation remained unaltered. SQUID magnetometry is a novel approach to quantify different iron forms in biological samples with high sensitivity that might provide new information for the understanding of the pathomechanisms of various diseased states as described previously [40,43]. The relative reduction of iron-containing compounds in Epi-treated vs. control rats in this study might be related to the inhibition of iron absorption in the enterocytes by Epi or its metabolites [56]. Indeed, there are animal and human studies that found reductions in iron bioavailability after the consumption of tea or various polyphenols-rich foods [57]. In our study, we showed the reduced levels of iron and/or iron-containing compounds in the blood of Epi-treated BHR. However, the exact mechanisms and chemical moiety of iron-containing compounds that are affected by Epi remain to be elucidated.

The lowered iron content level might be also involved in the reduced vascular ROS production. In our study, Epi significantly reduced the $O_2\bullet^-$ production in the aorta. This antioxidant effect was seen to partially persist after the cessation of the treatment, as the $O_2\bullet^-$ production in the aorta of the PE group was similar to that of the Epi group. The Epi-reduced $O_2\bullet^-$ production in the aorta of BHR was similar to our previous findings in the young SHR [39]. The antioxidant effects of Epi may also be a result of the direct scavenging of ROS [58], activation of antioxidant enzymes [59], or inhibition of pro-oxidant enzymes [58,60]. The suppression of the oxidative damage caused by Epi was also observed in the SHR [30] and other experimental models of hypertension [27–29].

It is well known that NO plays a key role in the modulation of BP [21]. In this study, the Epi stimulated NO production in the aorta and LHV (in contrast to the brain regions) of the young BHR, which, however, did not persist post treatment. The increased NO production due to Epi was found in various experimental models of hypertension: young SHR [30], adult SHR [31], as well as L-NAME-induced and fructose-induced hypertension [27,28] models. The observed NOS-stimulating effect of Epi is also supported by human studies in which the long-term administration of dark chocolate led to an increase of NO metabolites in blood in prehypertensive and hypertensive probands [52,61].

An important aim of our study was to investigate the genomic effects of the Epi during the stimulation of NO release and antioxidant defence in the LHV and aorta. The mRNA of all eNOS, iNOS and nNOS was present in the LHV and aorta in BHR. There are studies that found a significant role of nNOS-derived NO in local physiologic regulation of vascular tone [62], as well as in the physiological regulation of basal systemic vascular resistance and BP in healthy humans [63], despite the well-known role of eNOS-produced NO in flow-mediated dilatation. We found that the Epi did not alter the gene expression of the individual NOS isoforms and transcription factor *Nrf2* in the LHV and aorta. This suggests that the antioxidant effects of Epi in the abovementioned tissues are not related to the improved gene expressions of Nrf2-activated antioxidant enzymes. In addition, Nrf2 is involved in the NO-mediated regulation of iron metabolism, as it upregulates ferritin and ferroportin transcriptions, which results in the reduction of intracellular iron content [64]. However, these mechanisms seem to be less plausible in the cardiovascular system, as the *Nrf2* gene expression was unaltered during this experiment. Furthermore, Epi does not seem to modify PPAR-γ-mediated lipid metabolism and energy balance in the LVH and aorta [65] as *PPAR-γ* gene expressions were also found unaltered. There were significant correlations between *Nrf2* and *PPAR-γ*, as well as the individual NOS isoforms, despite the unchanged gene transcriptions, which suggests the regulatory role of Nrf2 in NOS expressions in the heart and aorta; this mechanism remained unaltered by Epi. The coordinating role of *Nrf2* in *PPAR-γ* expressions and endothelial NO generation was previously found by Luo et al. [66]. Regarding Epi effects, relatively few in vivo and in vitro studies have investigated the genomic and proteomic effects of purified Epi, especially the expressions of the genes that were investigated in this study. Our previous study showed that the expressions of all the NOS isoforms in the LHV of young SHR were unchanged [30], which was similar to those that were found in the young BHR. Mohamed et al. showed that *iNOS* gene expression was unaltered by Epi treatment in the brain of Wistar rats [67]. A study by Gómez-Guzmán et al. showed increased gene and protein expressions of Nrf2 upon Epi administration in the aorta of Wistar rats, which is in contrast to our findings [29]. Another study found the eNOS protein expression to be unchanged in Sprague-Dawley rats simultaneously treated with L-NAME and Epi [68]. Prince et al. investigated the alteration of the eNOS protein expression by administering Epi in the kidneys of Sprague-Dawley rats [69]. Rats that were simultaneously fed with fructose and Epi had similar expression levels as those of fructose-fed rats, which indicated that Epi does not affect the eNOS expression in the kidney [68]. Furthermore, Gómez-Guzmán et al. found the eNOS protein expression to be unaltered by Epi in the aorta of Wistar rats [70]. Thus, the BP-lowering effect of Epi was probably related to the increase of NOS activity and/or NO bioavailability due to improved redox conditions in the cardiovascular system of young BHR, similarly as it was found in SHR [30], without genomic effects being involved in the stimulation of NO release or antioxidant defence.

5. Conclusions

Our study showed that a two-week oral treatment of young BHR with Epi had a BP-lowering effect that persisted for two weeks after the cessation of the treatment. This was associated with a delayed reduction of anxiety-like behaviour. The mechanism underlying the BP-lowering effect of Epi was associated with the reduced relative content of the iron-containing compounds in the blood, reduced $O_2\bullet^-$ production, and stimulation of NOS activity in the LHV and aorta without increasing the mRNA expressions of the individual NOS isoforms or *Nrf2* and *PPAR-γ* transcription factors. In addition, our results suggest that continuous Epi treatment is required for maintaining elevated NO synthase activity and redox balance in the heart and aorta of young BHR. However, caution is needed due to the possible reduction of blood iron content with a long-term Epi intake.

Author Contributions: Conceptualization, I.B.; Data curation, M.K. and I.B.; Formal analysis, M.K. and I.B.; Funding acquisition, I.B.; Investigation, M.K., P.B., M.S., J.M. and I.B.; Methodology, M.S., J.M. and I.B.; Project administration, I.B.; Resources, I.B.; Visualization, M.K. and I.B.; Writing—original draft, M.K.; Writing—review & editing, M.S., J.M. and I.B. All authors have read and agreed to the published version of the manuscript.

Funding: This study was supported by the Scientific Grant Agency of the Ministry of Education, Science, Research and Sport of the Slovak Republic [grant No. 2/0160/17] and the Slovak Research and Development Agency [grant No. APVV-16-0263].

Acknowledgments: The authors thank Jana Petova and Bronislava Bolgacova for their technical assistance.

Conflicts of Interest: The authors have no conflicts of interest regarding the contents of the submission. The funders had no role in the design of the study; in the collection, analyses, or interpretation of data; in the writing of the manuscript, or in the decision to publish the results.

References

1. Chobanian:, A.V.; Bakris, G.L.; Black, H.R.; Cushman, W.C.; Green, L.A.; Izzo, J.L., Jr.; Jones, D.W.; Materson, B.J.; Oparil, S.; Wright, J.T., Jr.; et al. Seventh report of the Joint National Committee on Prevention, Detection, Evaluation, and Treatment of High Blood Pressure. *Hypertension* **2003**, *42*, 1206–1252. [CrossRef] [PubMed]

2. National Institutes of Health, National Heart, Lung, and Blood Institute, USA. 2012; Morbidity & Mortality: 2012 Chart Book on Cardiovascular, Lung, and Blood Diseases. Available online: https://www.nhlbi.nih.gov/files/docs/research/2012_ChartBook_508.pdf (accessed on 5 January 2020).

3. Redwine, K.M.; Daniels, S.R. Prehypertension in adolescents: Risk and progression. *J. Clin. Hypertens.* **2012**, *6*, 360–364. [CrossRef] [PubMed]

4. Assadi, F. Prehypertension: A warning sign of future cardiovascular risk. *Int. J. Prev. Med.* **2014**, *5* (Suppl. 1), S4–S9. [PubMed]

5. Huang, Y.; Wang, S.; Cai, X.; Mai, W.; Hu, Y.; Tang, H.; Xu, D. Prehypertension and incidence of cardiovascular disease: A meta-analysis. *BMC Med.* **2013**, *11*, 177. [CrossRef]

6. Williams, B.; Mancia, G.; Spiering, W.; Agabiti Rosei, E.; Azizi, M.; Burnier, M.; Clement, D.L.; Coca, A.; de Simone, G.; Dominiczak, A.; et al. 2018 ESC/ESH guidelines for the management of arterial hypertension. *Eur. Heart J.* **2018**, *39*, 3021–3104. [CrossRef]

7. Whelton, P.K.; Carey, R.M.; Aronow, W.S.; Casey, D.E., Jr.; Collins, K.J.; Dennison Himmelfarb, C.; DePalma, S.M.; Gidding, S.; Jamerson, K.A.; Jones, D.W.; et al. 2017 ACC/AHA/AAPA/ABC/ACPM/AGS/APhA/ASH/ASPC/NMA/PCNA Guideline for the prevention, detection, evaluation, and management of high blood pressure in adults: Executive summary: A report of the American college of cardiology/American heart association task force on clinical practice guidelines. *Circulation* **2018**, *138*, e426–e483. [CrossRef]

8. Baumann, M.; Janssen, B.J.; Hermans, J.J.; Peutz-Kootstra, C.; Witzke, O.; Smits, J.F.; Struijker Boudier, H.A. Transient AT1 receptor-inhibition in prehypertensive spontaneously hypertensive rats results in maintained cardiac protection until advanced age. *J. Hypertens.* **2007**, *25*, 207–215. [CrossRef]

9. Baumann, M.; Megens, R.; Bartholome, R.; Dolff, S.; van Zandvoort, M.A.; Smits, J.F.; Struijker-Boudier, H.A.; De Mey, J.G. Prehypertensive renin-angiotensin-aldosterone system blockade in spontaneously hypertensive rats ameliorates the loss of long-term vascular function. *Hypertens. Res.* **2007**, *30*, 853–861. [CrossRef]

10. Dickhout, J.G.; Lee, R.M. Blood pressure and heart rate development in young spontaneously hypertensive rats. *Am. J. Physiol.* **1998**, *274*, H794–H800. [CrossRef]

11. Puzserova, A.; Kopincova, J.; Slezak, P.; Balis, P.; Bernatova, I. Endothelial dysfunction in femoral artery of the hypertensive rats is nitric oxide independent. *Physiol. Res.* **2013**, *62*, 615–629.

12. Sarenac, O.; Lozic, M.; Drakulic, S.; Bajic, D.; Paton, J.F.; Murphy, D.; Japundzic-Zigon, N. Autonomic mechanisms underpinning the stress response in borderline hypertensive rats. *Exp. Physiol.* **2011**, *96*, 574–589. [CrossRef] [PubMed]

13. Thiyagarajan, R.; Pal, P.; Pal, G.K.; Subramanian, S.K.; Bobby, Z.; Das, A.K.; Trakroo, M. Cardiovagal modulation, oxidative stress, and cardiovascular risk factors in prehypertensive subjects: Cross-sectional study. *Am. J. Hypertens.* **2013**, *26*, 850–857. [CrossRef] [PubMed]

14. Chrysohoou, C.; Panagiotakos, D.B.; Pitsavos, C.; Skoumas, J.; Economou, M.; Papadimitriou, L.; Stefanadis, C. The association between pre-hypertension status and oxidative stress markers related to atherosclerotic disease: The ATTICA study. *Atherosclerosis* **2007**, *192*, 169–176. [CrossRef] [PubMed]

15. Outten, F.W.; Theil, E.C. Iron-based redox switches in biology. *Antioxid. Redox Signal.* **2009**, *11*, 1029–1046. [CrossRef]

16. Marques, V.B.; Nascimento, T.B.; Ribeiro, R.F., Jr.; Broseghini-Filho, G.B.; Rossi, E.M.; Graceli, J.B.; dos Santos, L. Chronic iron overload in rats increases vascular reactivity by increasing oxidative stress and reducing nitric oxide bioavailability. *Life Sci.* **2015**, *143*, 89–97. [CrossRef]

17. Atsma, F.; Veldhuizen, I.; de Kort, W.; van Kraaij, M.; Pasker-de Jong, P.; Deinum, J. Hemoglobin level is positively associated with blood pressure in a large cohort of healthy individuals. *Hypertension* **2012**, *60*, 936–941. [CrossRef]

18. Zhu, Y.; Chen, G.; Bo, Y.; Liu, Y. Markers of iron status, blood pressure and incident hypertension among Chinese adults. *Nutr. Metab. Cardiovasc. Dis.* **2019**, *29*, 830–836. [CrossRef]

19. Senthil, S.; Krishndasa, S.N. Pre-hypertension in apparently healthy young adults: Incidence and influence of haemoglobin level. *J. Clin. Diagn. Res.* **2015**, *9*, CC10–CC12. [CrossRef]

20. Gaenzer, H.; Marschang, P.; Sturm, W.; Neumayr, G.; Vogel, W.; Patsch, J.; Weiss, G. Association between increased iron stores and impaired endothelial function in patients with hereditary hemochromatosis. *J. Am. Coll. Cardiol.* **2002**, *40*, 2189–2194. [CrossRef]

21. Bernatova, I. Endothelial dysfunction in experimental models of arterial hypertension: Cause or consequence? *BioMed Res. Int.* **2014**, *2014*, 598271. [CrossRef]

22. Forrester, S.J.; Booz, G.W.; Sigmund, C.D.; Coffman, T.M.; Kawai, T.; Rizzo, V.; Scalia, R.; Eguchi, S. Angiotensin II signal transduction: An update on mechanisms of physiology and pathophysiology. *Physiol. Rev.* **2018**, *98*, 1627–1738. [CrossRef] [PubMed]

23. Bernatova, I. Biological activities of (−)-epicatechin and (−)-epicatechin-containing foods: Focus on cardiovascular and neuropsychological health. *Biotechnol. Adv.* **2018**, *36*, 666–681. [CrossRef] [PubMed]

24. Ottaviani, J.I.; Borges, G.; Momma, T.Y.; Spencer, J.P.; Keen, C.L.; Crozier, A.; Schroeter, H. The metabolome of [2-^{14}C](−)-epicatechin in humans: Implications for the assessment of efficacy, safety, and mechanisms of action of polyphenolic bioactives. *Sci. Rep.* **2016**, *6*, 29034. [CrossRef] [PubMed]

25. Mayorga-Gross, A.L.; Esquivel, P. Impact of cocoa products intake on plasma and urine metabolites: A review of targeted and non-targeted studies in humans. *Nutrients* **2019**, *11*, 1163. [CrossRef]

26. Ellinger, S.; Reusch, A.; Stehle, P.; Helfrich, H.P. Epicatechin ingested via cocoa products reduces blood pressure in humans: A nonlinear regression model with a Bayesian approach. *Am. J. Clin. Nutr.* **2012**, *95*, 1365–1377. [CrossRef]

27. Piotrkowski, B.; Calabro, V.; Galleano, M.; Fraga, C.G. (−)-Epicatechin prevents alterations in the metabolism of superoxide anion and nitric oxide in the hearts of L-NAME-treated rats. *Food Funct.* **2015**, *6*, 155–161. [CrossRef]

28. Litterio, M.C.; Vazquez Prieto, M.A.; Adamo, A.M.; Elesgaray, R.; Oteiza, P.I.; Galleano, M.; Fraga, C.G. (−)-Epicatechin reduces blood pressure increase in high-fructose-fed rats: Effects on the determinants of nitric oxide bioavailability. *J. Nutr. Biochem.* **2015**, *26*, 745–751. [CrossRef]

29. Gomez-Guzman, M.; Jimenez, R.; Sanchez, M.; Zarzuelo, M.J.; Galindo, P.; Quintela, A.M.; Lopez-Sepulveda, R.; Romero, M.; Tamargo, J.; Vargas, F.; et al. Epicatechin lowers blood pressure, restores endothelial function, and decreases oxidative stress and endothelin-1 and NADPH oxidase activity in DOCA-salt hypertension. *Free Radic. Biol. Med.* **2012**, *52*, 70–79. [CrossRef]

30. Kluknavsky, M.; Balis, P. (−)-Epicatechin prevents blood pressure increase and reduces locomotor hyperactivity in young spontaneously hypertensive rats. *Oxid. Med. Cell. Longev.* **2016**, *2016*, 6949020. [CrossRef]

31. Galleano, M.; Bernatova, I.; Puzserova, A.; Balis, P.; Sestakova, N.; Pechanova, O.; Fraga, C.G. (−)-Epicatechin reduces blood pressure and improves vasorelaxation in spontaneously hypertensive rats by NO-mediated mechanism. *IUBMB Life* **2013**, *65*, 710–715. [CrossRef]

32. Gao, L.; Zimmerman, M.C.; Biswal, S.; Zucker, I.H. Selective Nrf2 gene deletion in the rostral ventrolateral medulla evokes hypertension and sympathoexcitation in mice. *Hypertension* **2017**, *69*, 1198–1206. [CrossRef] [PubMed]

33. Barancik, M.; Gresova, L.; Bartekova, M.; Dovinova, I. Nrf2 as a key player of redox regulation in cardiovascular diseases. *Physiol. Res.* **2016**, *65* (Suppl. 1), S1–S10. [CrossRef] [PubMed]

34. Majzunova, M.; Dovinova, I.; Barancik, M.; Chan, J.Y. Redox signaling in pathophysiology of hypertension. *J. Biomed. Sci.* **2013**, *20*, 69. [CrossRef] [PubMed]

35. Ahmed, S.M.; Luo, L.; Namani, A.; Wang, X.J.; Tang, X. Nrf2 signaling pathway: Pivotal roles in inflammation. *Biochim. Biophys. Acta Mol. Basis Dis.* **2017**, *1863*, 585–597. [CrossRef] [PubMed]

36. Kvandova, M.; Barancik, M.; Balis, P.; Puzserova, A.; Majzunova, M.; Dovinova, I. The peroxisome proliferator-activated receptor gamma agonist pioglitazone improves nitric oxide availability, renin-angiotensin system and aberrant redox regulation in the kidney of pre-hypertensive rats. *J. Physiol. Pharmacol.* **2018**, *69*, 231–243. [CrossRef]

37. Silva-Palacios, A.; Konigsberg, M.; Zazueta, C. Nrf2 signaling and redox homeostasis in the aging heart: A potential target to prevent cardiovascular diseases? *Ageing Res. Rev.* **2016**, *26*, 81–95. [CrossRef]

38. Sugawara, A.; Uruno, A.; Kudo, M.; Matsuda, K.; Yang, C.W.; Ito, S. Effects of PPARγ on hypertension, atherosclerosis, and chronic kidney disease. *Endocr. J.* **2010**, *57*, 847–852. [CrossRef]

39. Puzserova, A.; Slezak, P.; Balis, P.; Bernatova, I. Long-term social stress induces nitric oxide-independent endothelial dysfunction in normotensive rats. *Stress* **2013**, *16*, 331–339. [CrossRef]

40. Kumar, P.; Bulk, M.; Webb, A.; van der Weerd, L.; Oosterkamp, T.H.; Huber, M.; Bossoni, L. A novel approach to quantify different iron forms in ex-vivo human brain tissue. *Sci. Rep.* **2016**, *6*, 38916. [CrossRef]

41. Zysler, R.D.; Lima, E., Jr.; Vasquez Mansilla, M.; Troiani, H.E.; Mojica Pisciotti, M.L.; Gurman, P.; Lamagna, A.; Colombo, L. A new quantitative method to determine the uptake of SPIONs in animal tissue and its application to determine the quantity of nanoparticles in the liver and lung of Balb-c mice exposed to the SPIONs. *J. Biomed. Nanotechnol.* **2013**, *9*, 142–145. [CrossRef]

42. Hashimoto, S.; Oda, T.; Yamada, K.; Takagi, M.; Enomoto, T.; Ohkohchi, N.; Takagi, T.; Kanamori, T.; Ikeda, H.; Yanagihara, H.; et al. The measurement of small magnetic signals from magnetic nanoparticles attached to the cell surface and surrounding living cells using a general-purpose SQUID magnetometer. *Phys. Med. Biol.* **2009**, *54*, 2571–2583. [CrossRef] [PubMed]

43. Janus, B.; Bućko, M.S.; Chrobak, A.; Wasilewski, J.; Zych, M. Magnetic characterization of human blood in the atherosclerotic process in coronary arteries. *J. Magn. Magnet. Mat.* **2011**, *323*, 479–485. [CrossRef]

44. Slezak, P.; Puzserova, A.; Balis, P.; Sestakova, N.; Majzunova, M. Genotype-related effect of crowding stress on blood pressure and vascular function in young female rats. *BioMed. Res. Int.* **2014**, *2014*, 413629. [CrossRef] [PubMed]

45. Sagvolden, T.; Metzger, M.A.; Schiorbeck, H.K.; Rugland, A.L.; Spinnangr, I.; Sagvolden, G. The spontaneously hypertensive rat (SHR) as an animal model of childhood hyperactivity (ADHD): Changed reactivity to reinforcers and to psychomotor stimulants. *Behav. Neural. Biol.* **1992**, *58*, 103–112. [CrossRef]

46. Yamada, T.; Yamada, Y.; Okano, Y.; Terashima, T.; Yokogoshi, H. Anxiolytic effects of short- and long-term administration of cacao mass on rat elevated T-maze test. *J. Nutr. Biochem.* **2009**, *20*, 948–955. [CrossRef] [PubMed]

47. Messaoudi, M.; Bisson, J.F.; Nejdi, A.; Rozan, P.; Javelot, H. Antidepressant-like effects of a cocoa polyphenolic extract in Wistar-Unilever rats. *Nutr. Neurosci.* **2008**, *11*, 269–276. [CrossRef]

48. Abd El Mohsen, M.M.; Kuhnle, G.; Rechner, A.R.; Schroeter, H.; Rose, S.; Jenner, P.; Rice-Evans, C.A. Uptake and metabolism of epicatechin and its access to the brain after oral ingestion. *Free Rad. Biol. Med.* **2002**, *33*, 1693–1702. [CrossRef]

49. Horvathova, M.; Zitnanova, I.; Kralovicova, Z.; Balis, P.; Puzserova, A.; Muchova, J.; Kluknavsky, M.; Durackova, Z.; Bernatova, I. Sex differences in the blood antioxidant defense system in juvenile rats with various genetic predispositions to hypertension. *Hypertens. Res.* **2016**, *39*, 64–69. [CrossRef]

50. Zicha, J.; Kunes, J. Ontogenetic aspects of hypertension development: Analysis in the rat. *Physiol. Rev.* **1999**, *79*, 1227–1282. [CrossRef]

51. Petyaev, I.M.; Dovgalevsky, P.Y.; Chalyk, N.E.; Klochkov, V.; Kyle, N.H. Reduction in blood pressure and serum lipids by lycosome formulation of dark chocolate and lycopene in prehypertension. *Food Sci. Nutr.* **2014**, *2*, 744–750. [CrossRef]

52. Sudarma, V.; Sukmaniah, S.; Siregar, P. Effect of dark chocolate on nitric oxide serum levels and blood pressure in prehypertension subjects. *Acta Med. Indones.* **2011**, *43*, 224–228. [PubMed]

53. Ried, K.; Frank, O.R.; Stocks, N.P. Dark chocolate or tomato extract for prehypertension: A randomised controlled trial. *BMC Complement. Altern. Med.* **2009**, *9*, 22. [CrossRef] [PubMed]

54. Stroh, A.; Zimmer, C.; Gutzeit, C.; Jakstadt, M.; Marschinke, F.; Jung, T.; Pilgrimm, H.; Grune, T. Iron oxide particles for molecular magnetic resonance imaging cause transient oxidative stress in rat macrophages. *Free Rad. Biol. Med.* **2004**, *36*, 976–984. [CrossRef] [PubMed]

55. Dal-Pizzol, F.; Klamt, F.; Frota, M.L., Jr.; Andrades, M.E.; Caregnato, F.F.; Vianna, M.M.; Schroder, N.; Quevedo, J.; Izquierdo, I.; Archer, T.; et al. Neonatal iron exposure induces oxidative stress in adult Wistar rat. *Brain Res. Dev. Brain Res.* **2001**, *130*, 109–114. [CrossRef]

56. Mladenka, P.; Macakova, K.; Filipsky, T.; Zatloukalova, L.; Jahodar, L.; Bovicelli, P.; Silvestri, I.P.; Hrdina, R.; Saso, L. In vitro analysis of iron chelating activity of flavonoids. *J. Inorg. Biochem.* **2011**, *105*, 693–701. [CrossRef] [PubMed]

57. Delimont, N.M.; Haub, M.D.; Lindshield, B.L. The impact of tannin consumption on iron bioavailability and status: A narrative review. *Curr. Dev. Nutr.* **2017**, *1*, 1–12. [CrossRef]

58. Ruijters, E.J.; Weseler, A.R.; Kicken, C.; Haenen, G.R.; Bast, A. The flavanol (−)-epicatechin and its metabolites protect against oxidative stress in primary endothelial cells via a direct antioxidant effect. *Eur. J. Pharmacol.* **2013**, *715*, 147–153. [CrossRef]

59. Quine, S.D.; Raghu, P.S. Effects of (−)-epicatechin, a flavonoid on lipid peroxidation and antioxidants in streptozotocin-induced diabetic liver, kidney and heart. *Pharmacol. Rep.* **2005**, *57*, 610–615.

60. Steffen, Y.; Gruber, C.; Schewe, T.; Sies, H. Mono-*O*-methylated flavanols and other flavonoids as inhibitors of endothelial NADPH oxidase. *Arch. Biochem. Biophys.* **2008**, *469*, 209–219. [CrossRef] [PubMed]

61. Taubert, D.; Roesen, R.; Lehmann, C.; Jung, N.; Schomig, E. Effects of low habitual cocoa intake on blood pressure and bioactive nitric oxide: A randomized controlled trial. *JAMA* **2007**, *298*, 49–60. [CrossRef]

62. Melikian, N.; Seddon, M.D.; Casadei, B.; Chowienczyk, P.J.; Shah, A.M. Neuronal nitric oxide synthase and human vascular regulation. *Trends Cardiovasc. Med.* **2009**, *9*, 256–262. [CrossRef]

63. Shabeeh, H.; Khan, S.; Jiang, B.; Brett, S.; Melikian, N.; Casadei, B.; Chowienczyk, P.J.; Shah, A.M. Blood pressure in healthy humans is regulated by neuronal NO synthase. *Hypertension* **2017**, *69*, 970–976. [CrossRef] [PubMed]

64. Koskenkorva-Frank, T.S.; Weiss, G.; Koppenol, W.H.; Burckhardt, S. The complex interplay of iron metabolism, reactive oxygen species, and reactive nitrogen species: Insights into the potential of various iron therapies to induce oxidative and nitrosative stress. *Free Rad. Biol. Med.* **2013**, *65*, 1174–1194. [CrossRef] [PubMed]

65. Strand, E.; Lysne, V. Short-term activation of peroxisome proliferator-activated receptors alpha and gamma induces tissue-specific effects on lipid metabolism and fatty acid composition in male Wistar rats. *PPAR Res.* **2019**, *2019*, 8047627. [CrossRef] [PubMed]

66. Luo, Z.; Aslam, S.; Welch, W.J.; Wilcox, C.S. Activation of nuclear factor erythroid 2-related factor 2 coordinates dimethylarginine dimethylaminohydrolase/PPAR-gamma/endothelial nitric oxide synthase pathways that enhance nitric oxide generation in human glomerular endothelial cells. *Hypertension* **2015**, *65*, 896–902. [CrossRef] [PubMed]

67. Mohamed, R.H.; Karam, R.A.; Amer, M.G. Epicatechin attenuates doxorubicin-induced brain toxicity: Critical role of TNF-alpha, iNOS and NF-kappaB. *Brain Res. Bull.* **2011**, *86*, 22–28. [CrossRef]

68. Litterio, M.C.; Jaggers, G.; Sagdicoglu Celep, G.; Adamo, A.M.; Costa, M.A.; Oteiza, P.I.; Fraga, C.G.; Galleano, M. Blood pressure-lowering effect of dietary (−)-epicatechin administration in L-NAME-treated rats is associated with restored nitric oxide levels. *Free Rad. Biol. Med.* **2012**, *53*, 1894–1902. [CrossRef]

69. Prince, P.D.; Lanzi, C.R.; Toblli, J.E.; Elesgaray, R.; Oteiza, P.I.; Fraga, C.G.; Galleano, M. Dietary (−)-epicatechin mitigates oxidative stress, NO metabolism alterations, and inflammation in renal cortex from fructose-fed rats. *Free Rad. Biol. Med.* **2016**, *90*, 35–46. [CrossRef]

70. Gomez-Guzman, M.; Jimenez, R.; Sanchez, M.; Romero, M.; O'Valle, F.; Lopez-Sepulveda, R.; Quintela, A.M.; Galindo, P.; Zarzuelo, M.J.; Bailon, E.; et al. Chronic (−)-epicatechin improves vascular oxidative and inflammatory status but not hypertension in chronic nitric oxide-deficient rats. *Br. J. Nutr.* **2011**, *106*, 1337–1348. [CrossRef]

Article

Cocoa Flavonoids Reduce Inflammation and Oxidative Stress in a Myocardial Ischemia-Reperfusion Experimental Model

Sajeela Ahmed [1,†], Naseer Ahmed [1,2,†], Alessio Rungatscher [1,*], Daniele Linardi [1], Bibi Kulsoom [3], Giulio Innamorati [1], Sultan Ayoub Meo [4], Mebratu Alebachew Gebrie [1], Romel Mani [1], Flavia Merigo [5], Flavia Guzzo [6] and Giuseppe Faggian [1]

[1] Department of Surgery, Division of Cardiac Surgery, University of Verona, 37129 Verona, Italy; sajeelaahmed2@gmail.com (S.A.); dr.naseer99@gmail.com (N.A.); danielelinardi@yahoo.it (D.L.); giulio.innamorati@univr.it (G.I.); mebanat@yahoo.com (M.A.G.); romel.mani@univr.it (R.M.); giuseppe.faggian@univr.it (G.F.)
[2] Department of Biological and Biomedical Sciences, Aga Khan University, 74800 Karachi, Pakistan
[3] Department of Biochemistry, Jinnah Medical & Dental College, 74800 Karachi, Pakistan; drknpk@yahoo.com
[4] Department of Physiology, College of Medicine, King Saud University, 11461 Riyadh, Saudi Arabia; smeo@ksu.edu.sa
[5] Department of Biomedicine, Division of Histology, University of Verona, 37134 Verona, Italy; Flavia.merigo@univr.it
[6] Department of Biotechnology, Division of Biology and Botany, University of Verona, 37134 Verona, Italy; flavia.guzzo@univr.it
* Correspondence: alessio.rungatscher@univr.it
† Both authors contributed equally and shared first authorship.

Received: 10 January 2020; Accepted: 16 February 2020; Published: 18 February 2020

Abstract: Consumption of flavonoid-rich nutraceuticals has been associated with a reduction in coronary events. The present study analyzed the effects of cocoa flavonols on myocardial injury following acute coronary ischemia-reperfusion (I/R). A commercially available cocoa extract was identified by chromatographic mass spectrometry. Nineteen different phenolic compounds were identified and 250 mg of flavan-3-ols (procyanidin) were isolated in 1 g of extract. Oral administration of cocoa extract in incremental doses from 5 mg/kg up to 25 mg/kg daily for 15 days in Sprague Dawley rats ($n = 30$) produced a corresponding increase of blood serum polyphenols and become constant after 15 mg/kg. Consequently, the selected dose (15 mg/kg) of cocoa extract was administered orally daily for 15 days in a treated group ($n = 10$) and an untreated group served as control ($n = 10$). Both groups underwent surgical occlusion of the left anterior descending coronary artery and reperfusion. Cocoa extract treatment significantly reversed membrane peroxidation, nitro-oxidative stress, and decreased inflammatory markers (IL-6 and NF-kB) caused by myocardial I/R injury and enhanced activation of both p-Akt and p-Erk1/2. Daily administration of cocoa extract in rats is protective against myocardial I/R injury and attenuate nitro-oxidative stress, inflammation, and mitigates myocardial apoptosis.

Keywords: flavonoids; cocoa extract; ischemia-reperfusion injury; oxidative stress; apoptosis; inflammatory markers

---

1. Introduction

In United States, about one million people suffer from myocardial infarction per year [1]. Ischemic heart disease provoked by limited blood supply to cardiomyocytes upon occlusion of coronary vessels is the main precursor of myocardial infarction. Prevention of tissue damage in patients with ischemic heart disease can be effectively accomplished through reperfusion of the ischemic myocardial tissue [2,3]. However, reperfusion itself is responsible for inducing injury to cardiomyocytes via multiple mechanisms, leading to heart failure [4,5]. Oxidative stress and myocardial inflammatory pathways play a leading role in the pathogenesis of myocardial ischemia-reperfusion (I/R) injury [6]. Oxidative stress is caused by an intracellular redox imbalance between pro- and anti-oxidants [7–9]. Exogenous antioxidants might influence the course of the ischemic heart disease by providing therapeutic substances which help in restoring and maintaining a balanced system [10]. Therefore, plant resources with anti-oxidant activity would be worthy natural substances for protection against ischemic heart disease. The progression of reperfusion, thus, is accompanied by the development of oxidative stress with the generation of free radicals and leukocyte activation that lead to myocytes apoptosis [11]. Exogenous antioxidants might influence the course of the ischemic heart disease by contributing to restoring a balanced system.

Many scientists have been working to explore antioxidant-rich natural sources to protect against myocardial I/R injury [12]. Many studies have reported plants as the main reservoir of natural antioxidants and anti-inflammatory compounds, thus pointing towards the protective role of plant products against inflammatory and reperfusion injury [12–15].

Cocoa and chocolate products contribute high levels of flavonoids among commonly consumed foods and have been historically used as a medicine to cure inflammation, pain, and numerous other diseases [13,16]. It is observed that adipose tissue inflammation can be reduced by long-term cocoa supplementation. In a recent study reported by Akinmoladun et al., antioxidant-containing extracts of cocoa and the kola nut tree have shown a protective effect against myocardial I/R injury using Langerdorff-perfused rat hearts [13]. In another study, flavonoids (5-hydroxy derivatives: 5-hydroxy flavone, apigenin, chrysin, and naringenin) lowered myocardial tissue injury and improved post-ischemic functional recovery [17]. Earlier, an in vitro study reported reduced IL-1 mRNA expression and IL-2 secretion by T-cells in polyphenol containing cocoa liquor [18]. However, previous studies lack biochemical analysis of inflammatory markers and signaling protein activities, and evaluation of myocardial nuclei apoptotic levels in rat hearts after I/R injury. Moreover, to the best of our knowledge, dose optimization of the cocoa extract in blood sera in rats has not been documented previously in literature.

In this study, we investigate the effects of a commercially available cocoa extract on oxidative stress, inflammation, and apoptosis to ascertain whether it can protect myocardium in an in vivo experimental model of I/R injury.

2. Methods and Materials

2.1. Chemicals and Materials Used

CocoaVia® was purchased from Mars Inc., (Hackettstown, NJ, USA). Primary antibodies Nitro-tyrosine, IL-6, NFkB2, P-Erk, P-akt antibody were obtained from Bioss antibodies, Novusbio, and Cell Signaling Technology (Danvers, MA USA). In Situ Cell Death Detection Kit, AP was from Roche (Basel, Switzerland).

2.2. Animals

Thirty healthy Sprague Dawley (SD) male rats of average weight 300–350g were used for the dose response experiment and 20 rats were used to assess the effect of cocoa extract on ischemia reperfusion injury. All animal experiments were done per the ethical guidelines reviewed and approved by "University of Verona Ethical Committee and the Italian Ministry of Health (341/2016-PR) at C.I.R.S.A.L. (Interdepartmental Research Centre for Laboratory Animals) of the Biological Institutes, University of Verona, and Verona, Italy".

2.3. Extraction and Identification of CocoaVia® Contents

Total phenolic compounds of the cocoa extract supplement (Cocoavia; Mars Inc., Hackettstown, NJ, USA) made by patented process (Cocoapro; Mars Inc., Hackettstown, NJ, USA) were extracted with three volumes of ice-cold methanol (*w/v*)). Samples were mixed, sonicated (15 min, 4 °C), centrifuged (16,000× *g* rpm, 10 min, 4 °C), and filtered through 0.2 μm pore filters. The supernatant samples were diluted 1:100 and 1:10 (*v/v*) with LC-MS grade methanol for HPLC-ESI-MS. The samples were further diluted 1:2 with LC-MS grade water and passed through Minisart 0.2 μm filters following protocol [19]. Each sample was analyzed in two technical and three biological replicates, with 20 μL injection volumes. The unknown metabolites of samples were analyzed using HPLC-ESI-MS (Esquire 6000, Bruker Daltonics, Billerica, USA), followed by ESI (Electrospray Ionization)-base peaks, MS/MS, and MS3 fragmentation at its retention time and mass to charge ratio (*m/z*). For this purpose, an "in house" library of commercial standard spectra, scientific literature, and online databases such as Mass bank (www.massbank.jp) were used. The quantification of the metabolites in the methanolic extracts was carried out through HPLC-DAD (Beckman Coulter Gold 126 Solvent Module coupled with a Gold 168 Diode Array Detector), relying on the calibration curves of authentic standard compounds.

2.4. Quantification of Polyphenols Levels in Blood Samples

Thirty rats were divided into six groups of five rats each. Among them, group 1 served as control, fed with normal rat diet, while the animals of other groups were administrated oral gavage with 5, 10, 15, 20, and 25 mg/kg body weight of cocoa extract dissolved in water, five times per week for 15 days, respectively. All animals were kept at a temperature of 22–24 °C and fed with a regular pellet diet ad libitum.

A total of 1 mL of blood was collected, 1 hour after oral administration of cocoa extract powder from each rat, which showed a higher concentration of flavonoids in plasma between 30 and 60 min through lateral tail vein [19]. Blood was deposited in clean heparinized glass tubes, centrifuged (3500 rpm for 15 min), and stored at −80 °C for the HPLC-ESI-MS analysis. The total amount of flavonoid in blood sera from each sample were calculated by HPLC-ESI-MS [19,20].

2.5. Induction of Ischemia/Reperfusion Injury and Tissue Collection

Twenty rats were divided into two groups, control and treated, with ten (10) in each group. The selected dose of 15 mg/kg of cocoa from our previous results was given to the treated group ($n = 10$) once a day for 15 days. The control group remained untreated. Rats ($n = 20$) were anesthetized with 5% isoflurane in 50% O_2 administered through a facial mask and maintained anesthesia using 2% isoflurane throughout procedure. These anesthetized animals were subjected to ischemia for 30 min by left anterior descending (LAD) coronary artery ligation, followed by 120 min reperfusion, as previously described [21]. The onset of ischemia was confirmed when cyanosis developed on the wall of the ischemic myocardium, as evidenced by saddleback-type (ST) segment elevation and a significant T-wave increase recorded on the electrocardiograph Power Lab data acquisition system (model ML866, AD Instruments, Colorado Springs, CO) and Animal Bio Amp (model ML136, AD Instruments)). Then, the ligation was opened to allow reperfusion to the ischemic part of the myocardium for 120 min. Hearts were excised as demonstrated in Figure 1, the ischemic part of the left ventricle was split into two: one portion was stored in 4% formalin for immunohistochemistry and Terminal Deoxynucleotidyl Transferase dUTP Nick End Labeling (TUNEL) analysis, the second portion was stored in liquid nitrogen for of nitro-oxidative stress measurements. Experimental design is illustrated in Figure 1.

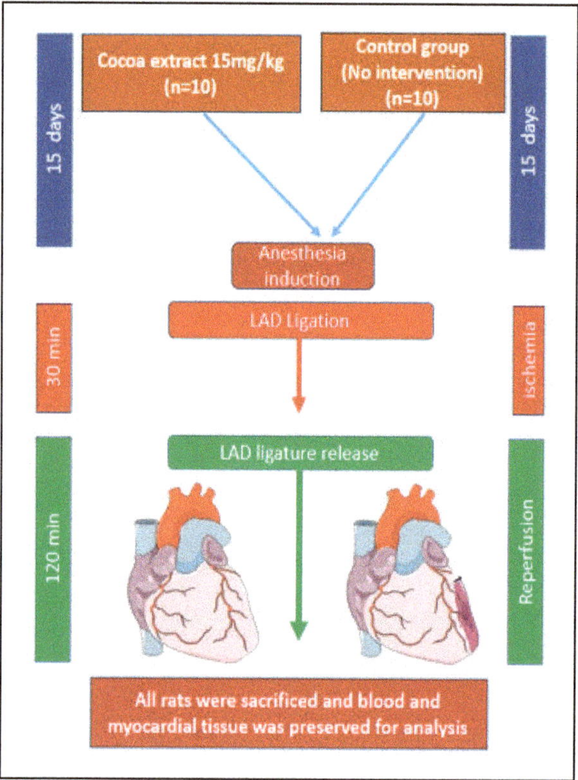

Figure 1. Schematic experimental design.

2.6. Immunohistochemistry Staining

To access the inflammation and nitrative stress and signaling pathways activation in myocardial tissue, samples (n = 10 each group) were embedded in paraffin. Fixed tissues were sectioned 3 μm, deparaffinized, and dehydrated with two grades of xylene and four grades of ethanol, following the method in [22]. After antigen retrieval and endogenous peroxidase activity procedure, all sections were incubated with primary antibodies, IL-6 (1:100), NF-κB2 (1:500), Nitrotyrosine (1:300), p-Erk (1:200), and p-Akt (1:200) (Sigma–Aldrich, United Kingdom), diluted in antibody diluent and kept at 4 °C for overnight. Following the incubation with primary antibodies, tissue sections were rinsed, and incubated with the biotinylated anti-rabbit secondary antibody (1:400), avidin–biotin complex substrate, and diaminobenzidine (Dako Corp., Carpinteria, CA, USA). Sections were rinsed two times and mounted after dehydration in three grades of ethanol and cleared in two grades of xylene. The negative control was used to confirm and check the absence of the signal or specificity of staining. Image acquisition of all the sections was done under Olympus System BX51 Universal research microscopy (Olympus corporation, Tokyo, Japan). The images were analyzed by using ImageJ software (NIH, Bethesda, MD, USA) to quantify the strength of immune-peroxidase staining in heart tissue.

2.7. Oxidative Stress Measurement

The analysis of malondialdehyde (MDA) levels in myocardial tissue samples collected from the affected area was performed as described by Ben Maunsour et al. [23].

The reactive oxygen species (ROS) in myocardial tissue was measured using a spectrophotometry and the absorbance determined (peak at 505 nm max) value is expressed as Carr. Units, as explained by Rizzo et al. [24].

2.8. Terminal Deoxynucleotidyl Transferase dUTP Nick End Labeling (TUNEL) Assay

TUNEL assay is a method that labels 3′-hydroxyl termini in the double-stranded DNA fragments generated as a result of apoptosis. Paraffin-mounted heart sections from control and treated groups ($n = 10$) were deparaffinized with xylene and descending concentrations of ethanol. Apoptotic cells were stained following the protocol provided by the In Situ Cell Death Detection Kit, AP (ref. 11684809910, Version 11, Roche Diagnostics, Indianapolis, IN, USA). Hoechst 33342 solution was used for staining nuclei and examined under confocal microscope. The images were analyzed by ImageJ software (NIH, Bethesda, MD, USA), average mean fluorescent intensities (MFIs) were measured by multiplying the intensities of each image [25].

2.9. Statistical Analysis

Analyses were performed using SPSS software version 21 (SPSS Inc., Chicago, IL, USA). Comparison between control and experimental groups was statistically evaluated by the Students t-test, and p-value < 0.05 was considered statistically significant. All the results were expressed as mean ± SD.

3. Results

3.1. Identification of Phenolic Compounds

A total of 250 mg of flavan-3-ols (procyanidin) was identified from 1 g of cocoa extract. From this, 19 different phenolic compounds were determined (Table 1 and Figure 2a).

Table 1. Phenolic molecules identified from the cocoa extract samples analyzed by chromatographic mass spectra in negative ion mode.

No.	(−) *m/z*	No.	(−) *m/z*
1	Di-hexose, Di-hexose (2M−H)	11	Procyanidin pentamer
2	N-[3′,4′-dihydroxy-(Z)-cinnamoyl]-ʟ-aspartic acid	12	Procyanidin monomer
3	Catechin hexose	13	Procyanidin aptamer
4	Catechin	14	Procyanidin hexamer
5	Procyanidin trimer	15	Procyanidin
6	Procyanidin tetramer	16	Quercetin hexose
7	Procyanidin B2	17	Procyanidin derivative
8	Epicatechin	18	Ellagic acid pentose
9	Procyanidin trimer, Catechin derivative	19&20	(Epi) catechin ethyl trimmers
10	Procyanidin tetramer		

3.2. Dose Response and Dose Adjustment

Oral administration of cocoa extract in incremental doses from 5 mg/kg up to 25 mg/kg daily for 15 days produced a corresponding increase in blood serum polyphenols and became constant after 15 mg/kg (Figure 2B).

3.3. Anti-Inflammatory Effect of Cocoa

The extent of inflammation after I/R was measured by the expression of IL-6 and NF-kB in the myocardium. Immunohistochemistry revealed a reduced expression of IL-6 (Figure 3A,B) was significantly lowered after cocoa extract treatment 15 mg/kg ($p < 0.01$) in the ischemia-reperfusion injury model. The levels of NF-kB were remarkably low in the treated group as compared to control (Figure 3C,D).

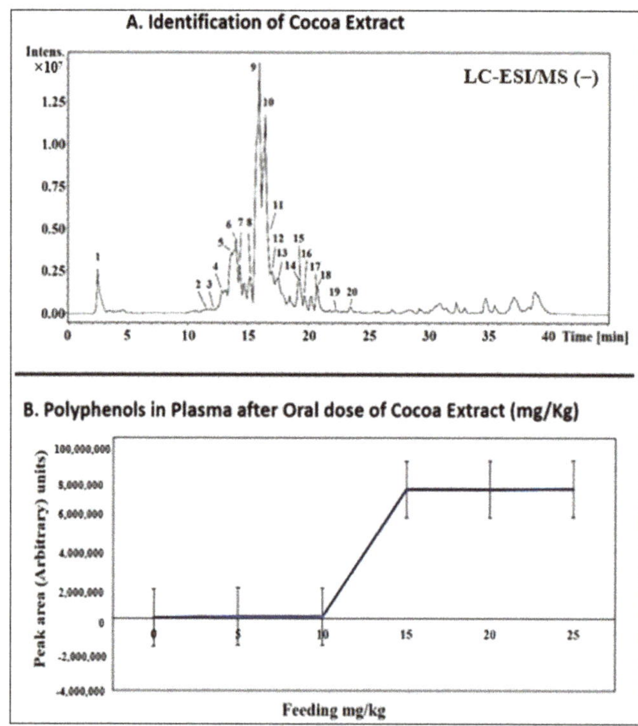

Figure 2. Cocoa Extract: (**A**) shows HPLC-ESI-MS graph with various peaks correlating with 19 different phenolic compounds found in cocoa extract. (**B**) shows a curve correlating different oral doses (mg/kg) of cocoa extract and blood levels of phenolic compounds in rats. It shows an optimal dose of 15 mg/kg to be the optimal dose of cocoa extract (six groups ($n = 30$) and each group ($n = 5$)).

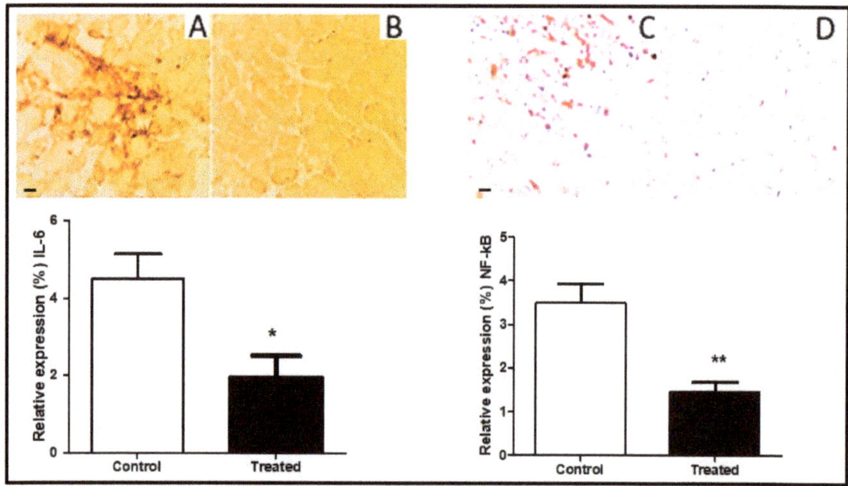

Figure 3. Immunohistochemical analysis showing expression (representative images, 20× magnification) of IL-6 and NF-kB2 in myocardial tissue of cocoa treated rats ($n = 10$) compared to control rats ($n = 10$), (*$p = 0.0032$, **$p = 0.0001$ respectively). presented as mean ± SD and scale bar is 100 μm. (**A**,**B**): IL-6, (**C**,**D**): NF-kB.

3.4. Nitro-Oxidative Stress Attenuation with Cocoa

Oxidative stress was studied by identifying the levels of the "lipid peroxidation index, Thiobarbituric acid reactive substances (TBARS)". TBARS expressed as malondialdehyde (MDA) levels was lower in the treated group as compared to control (Figure 4A). The concentration of ROS was also attenuated in the treated group as compared to control (Figure 4B). Nitrosative stress was assessed by measuring nitrotyrosine levels in myocardial tissue. The nitrotyrosine was reduced by cocoa treatment ($p < 0.05$) (Figure 4C,D).

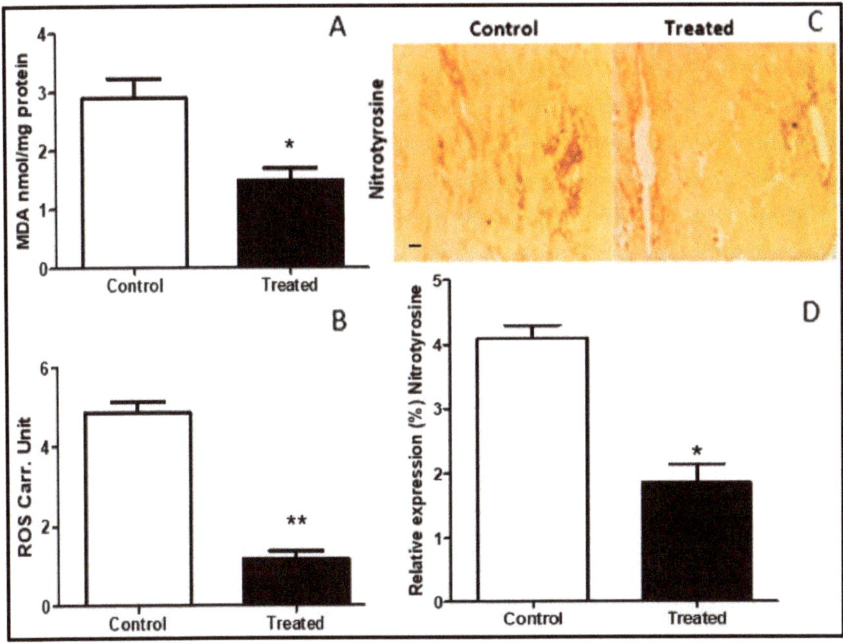

Figure 4. Effects of cocoa on oxidative stress on myocardial tissue rats: Lower oxidative stress was recorded in cocoa-treated rats ($n = 10$) myocardial tissue as compared to control ($n = 10$). (**A**) Lipid peroxidation index (TBARS) expressed as malondialdehyde (MDA) concentration measured in nmol/μg ($p = 0.031$), * $p < 0.05$. (**B**) Reactive oxygen species (ROS) measured as Carr. Unit (Carratelli Unit) ($p = 0.0001$), ** $p < 0.001$. (**C,D**) Expression of nitrotyrosine staining in myocardial tissue in the control and treated groups (20×) ($p = 0.04$), * $p < 0.05$. Data illustrated in the graphs are presented as mean ± SD and scale bar is 100 μm.

3.5. Akt and Erk1/2 Signaling Pathways Activation

Immuno-peroxidase analysis of p-Akt and p-ERK1/2 in myocardial tissue from rats treated with cocoa extract and control group was performed. Enhanced phosphorylation of Akt (Figure 5A,B,E) and ERK1/2 (Figure 5C,D,F) was observed as compared to control group ($p < 0.05$ in both cases).

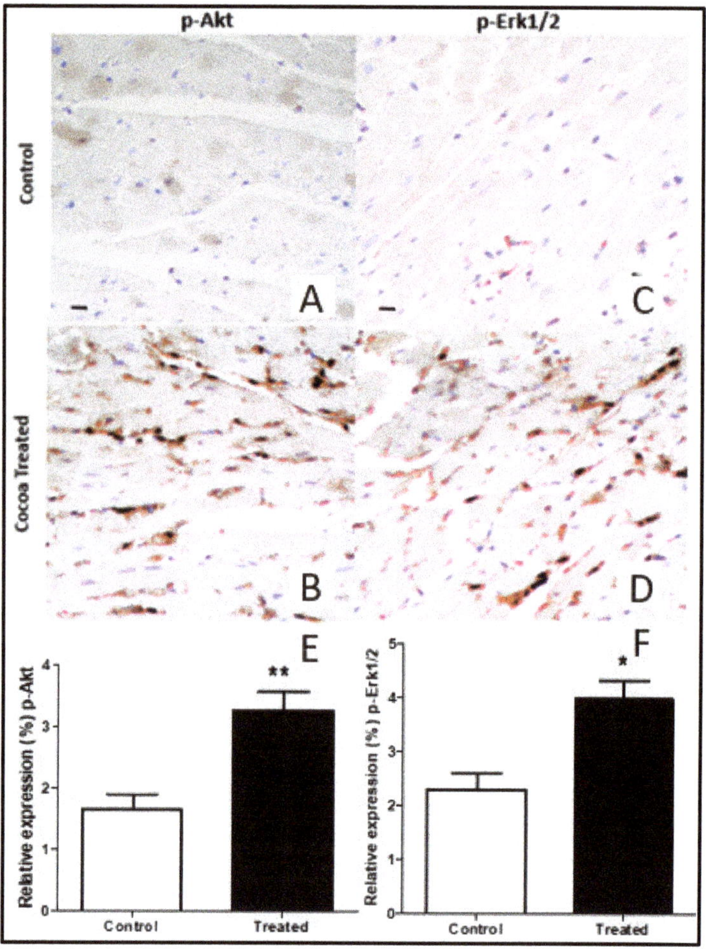

Figure 5. Immunohistochemical labeling of p-Akt (**A,B,E**) and p-Erk1/2 (**C,D,F**) in cocoa-treated rat myocardial tissue compared to control. The graphs show a significant elevation in the activation of p-Akt and p-Erk1/2 in the myocardial tissue of cocoa-treated rats ($n = 10$) compared to control ($n = 10$). Representative images (20×). Data illustrated in the bar graph are presented as mean ± SD. *p*-value less than 0.05 considered statistically significant. P-ERK1/2, phosphorylated extracellular signal regulated kinases $\frac{1}{2}$; p-Akt, phosphorylated serine-threonine protein kinase, (* $p < 0.05$, ** $p < 0.001$).

3.6. TUNEL Assay

TUNEL-positive nuclei were less in number in the treated group (A) as compared to the control group (B). Yellow arrows indicate apoptotic nuclei (Figure 6A–C). There was a significant reduction in apoptosis in treated samples as compared to control ($p < 0.001$).

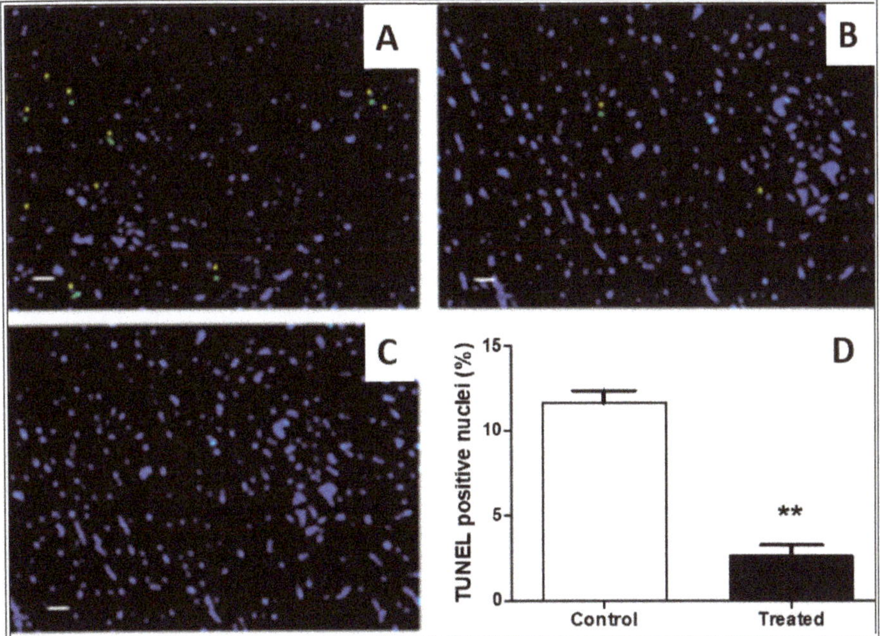

Figure 6. TUNEL Assay: (**A,B,C**): Representative images (*n* = 10 for each group) immunofluorescent staining for TUNEL-positive nuclei in control, treated, and negative control groups (20× magnification). TUNEL-positive myocytes were less in number in the treated group (**A**) as compared to control group (**B**). Yellow arrows indicate apoptotic nuclei and (**C**) negative control. *p*-value = 0.002 considered significant (** *p* < 0.001, control versus treated), scale bar is 100 μm. (**D**): Graphical presentation of apoptotic nuclei in control vs treated groups.

4. Discussion

Cocoa has the highest flavonol contents of all foodstuffs and its extract contains a considerable concentration of proanthocyanidins [16]. Flavonoids characterize a main division of phenolic compounds, and they are greatly active scavengers of most oxidizing molecules and free radicals involved in several diseases [17,18], including cardiovascular diseases.

In the present study, a total of 250 mg of procyanidins was determined in 1 gram of cocoa extract. From this, 19 different phenolic compounds were identified (Figure 1 and Table 1). Many studies have shown the protective effect of cocoa in myocardial ischemia-reperfusion injury and in improving post-ischemic functional recovery [13,17]. Thus, it was observed that cocoa extract has anti-inflammatory properties [26] and it can protect against myocardial injury [27]. Even though flavonoids have many health-related benefits, their bioavailability is a major concern [28]. We hypothesized that these phenolic compounds, introduced daily with cocoa extract supplement, might contribute to reducing inflammatory markers, oxidative stress, myocardial apoptosis, and activating pro-survival pathways in the heart exposed to I/R injury. In the present study, different concentrations of a commercially available cocoa extract were investigated in a dose-response assessment in order to establish the optimal dose in a daily administration regimen in order to have maximal plasma concentration of polyphenols. An optimal dose of 15 mg/kg/BW was administered daily to investigate the possible cardioprotective effects during acute myocardial I/R.

The expression of inflammatory mediators, which recruits leukocytes into inflammatory sites, was reduced in the cocoa-treated rat heart (Figure 3). During the inflammatory process at tissue, the usual trend is the elevation of IL-6 and NF-kB levels [29]. These mediators were significantly reduced in cocoa extract-treated rat hearts. In vitro reduction in other inflammatory meditators, such as IL-1 mRNA expression and IL-2 secretion by T-lymphocyte treated with cocoa liquor was also reported [17]. Thus, reduction in inflammatory markers in cocoa-treated rat hearts suggests an anti-inflammatory and cardioprotective role of cocoa extract flavonoids.

Malondialdehyde (MDA), a marker of oxidative stress, was attenuated in cocoa-treated heart tissue (Figure 4A). The levels of reactive oxygen species were found to be significantly higher in cocoa-treated rat myocardium as compared to the control group after the induction of I/R (Figure 4B). Similarly, Akinmoladun et al. explored the protective effects of cocoa and kola nut tree extracts, and also reported a reduction in oxidative stress and post I/R injury in an in-vitro model [13]. Nitric oxide (NO) produced by endothelial nitric oxide synthase (eNOS) constitutively is reduced in ischemia but inducible NOS is activated by I/R, thus increasing nitrosative stress. Nitrotyrosine (peroxynitrite) levels, which indicate nitrosative stress [30] in myocardial tissue due to I/R, were lower in cocoa-treated rat heart in our study (Figure 4C,D).

The ultimate step of myocardial injury in reperfusion injury is apoptosis, and inhibition of apoptotic pathways demonstrated significant cardioprotection. Reduced apoptosis leads to reduced myocardial injury [31]. Apoptosis could be due to mitochondrial dysfunction, which results in low energy for myocyte contraction. Additionally, there may be an increase in oxidative stress that directly damages and induces apoptosis in myocytes [32–34]. During I/R, the fuel preference switches to glucose. This alteration safeguards the heart, in part, because "free acids waste more oxygen to be oxidized". However, over time, while not oxidized, fatty acids hoard inside the myocytes, leading to cardiac lipotoxicity [35,36]. Nitrite reacts with critical thiols to form nitrosothiols, which act as antioxidants that prevent the irreversible oxidation of proteins and lipids during the early oxidative burst of reperfusion" [37,38].

Despite the pharmacological effectiveness of cardioprotective drugs, novel products which inhibit ischemic organ damage are required and care is being given to discovering novel pharmacological agents from plants [39]. Infact, the medicinal plants' antioxidant content may contribute to protection from diseases. Commonly found in plants, phenolic compounds are major antioxidant phytochemicals [27]. When the plants are consumed, these phytochemicals contribute to the intake of natural antioxidants in the diets of animals as well as humans.

The reduced levels of myocardial apoptosis observed in cocoa-treated rats demonstrate potential effect of flavonoids containing cocoa in reducing myocardial apoptosis. The toxicological study also indicated the contribution of epicatechin and catechin in cocoa in the reduction of apoptosis via the inhibition of amyloid-β protein [40]. The reduced myocardial apoptosis could be due to the antioxidant content of cocoa, which might contribute to the protection against oxidative stress produced by I/R.

In our study, immuno-peroxidase analysis of myocardial tissue from rats treated with cocoa extract showed elevation in the activity of p-Erk1/2 and p-Akt (Figure 5). Pro-survival pathways proteins (Erk1/2 and Akt) are known for their contribution to cell survival during ischemia-reperfusion injury. Enhanced phosphorylation of Erk1/2 and Akt were observed with cocoa treatment, which leads to cardioprotection [41]. The TUNEL assay, which is good indicator to measure apoptosis, demonstrated reduced apoptosis in cocoa-treated heart tissues as compared to control (Figure 6).

Therefore, this study suggests that treating rats with cocoa extract significantly attenuates inflammation, oxidative stress, and apoptosis in the myocardium and the process can be modulated by the activation of Erk1/2 and Akt pathways.

The present study investigated only one commercially available cocoa extract. Other products could give different results. Additionally, we investigated healthy rats; in order to translate the results to humans they should be confirmed in a randomized clinical study in patients with ischemic cardiomyopathy taking different medications.

5. Conclusions

Daily supplementation of cocoa extract attenuates myocardial I/R injury, limiting oxidative and nitrosidative stress and inflammation with a reduction in myocardial apoptosis.

Author Contributions: S.A., N.A.: experimental work, bench work, and manuscript writing; A.R.: Designed protocol, experimental work, data analysis, manuscript writing; G.I., M.A.G., F.M. and F.G.: Bench work; D.L. and R.M.: Experimental work; B.K. and S.A.M.: Manuscript writing and critical review; G.F.: Supervised and reviewed experimental protocol. All authors have read and agreed to the published version of the manuscript.

Funding: This research was funded by "Researchers Supporting Project Number (RSP-2019/47), King Saud University, Riyadh, Saudi Arabia, Cardiac Surgery Division, University of Verona, Verona, Italy and Aga Khan University, Karachi, Pakistan.

Acknowledgments: Authors are thankful to the King Saud University, Riyadh, Saudi Arabia for providing support through "Researchers Supporting Project (RSP-2019/47), King Saud University, Riyadh, Saudi Arabia". We would like to thank Claudia Fiorini and other staff of the surgical lab and anatomy pathology lab, Policlinico Borgo Roma, University of Verona for their continuous support and technical assistance in molecular analysis.

Conflicts of Interest: The authors declare no conflict of interest. The producer of cocoa extract has not been informed before, during and after the execution of the study.

References

1. Frank, A.; Bonney, M.; Bonney, S.; Weitzel, L.; Koeppen, M.; Eckle, T. Myocardial Ischemia Reperfusion Injury: From Basic Science to Clinical Bedside. *Semin. Cardiothorac. Vasc. Anesth.* **2012**, *16*, 123–132. [CrossRef] [PubMed]

2. Li, X.; Liu, M.; Sun, R.; Zeng, Y.; Chen, S.; Zhang, P. Protective Approaches against Myocardial Ischemia Reperfusion Injury. *Exp. Ther. Med.* **2016**, *12*, 3823–3829. [CrossRef] [PubMed]

3. Kuznetsov, A.; Javadov, S.; Margreiter, R.; Grimm, M.; Hagenbuchner, J.; Ausserlechner, M.J. The Role of Mitochondria in the Mechanisms of Cardiac Ischemia-Reperfusion Injury. *Antioxidants* **2019**, *8*, 454. [CrossRef] [PubMed]

4. Ahmed, N.; Mehmood, A.; Linardi, D.; Sadiq, S.; Tessari, M.; Meo, S.A.; Rehman, R.; Hajjar, W.M.; Muhammad, N.; Iqbal, M.P.; et al. Cardioprotective Effects of Sphingosine-1-Phosphate Receptor Immunomodulator FTY720 in a Clinically Relevant Model of Cardioplegic Arrest and Cardiopulmonary Bypass. *Front. Pharmacol.* **2019**, *10*, 802. [CrossRef] [PubMed]

5. Yellon, D.M.; Hausenloy, D.J. Myocardial Reperfusion Injury. *New Engl. J. Med.* **2007**, *357*, 1121–1135. [CrossRef] [PubMed]

6. Timmers, L.; Pasterkamp, G.; De Hoog, V.C.; Arslan, F.; Appelman, Y.; De Kleijn, D.P.V. The Innate Immune Response in Reperfused Myocardium. *Cardiovasc. Res.* **2012**, *94*, 276–283. [CrossRef]

7. Gulcin, I.; Oktay, M.; Küfrevioğlu, O.I.; Aslan, A. Determination of Antioxidant Activity of Lichen Cetraria Islandica (L) Ach. *J. Ethnopharmacol.* **2002**, *79*, 325–329. [CrossRef]

8. Zhao, D.; Yang, J.; Yang, L. Insights for Oxidative Stress and mTOR Signaling in Myocardial Ischemia/Reperfusion Injury under Diabetes. *Oxidative Med. Cell. Longev.* **2017**, *2017*, 1–12. [CrossRef]

9. Habtemariam, S. Modulation of Reactive Oxygen Species in Health and Disease. *Antioxidants* **2019**, *8*, 513. [CrossRef]

10. Branen, A.L. Toxicology and Biochemistry of Butylated Hydroxyanisole and Butylated Hydroxytoluene. *J. Am. Oil Chem. Soc.* **1975**, *52*, 59–63. [CrossRef]

11. Giordano, F.J. Oxygen, Oxidative Stress, Hypoxia, and Heart Failure. *J. Clin. Invest.* **2005**, *115*, 500–508. [CrossRef] [PubMed]

12. Forte, M.; Conti, V.; Damato, A.; Ambrosio, M.; Puca, A.A.; Sciarretta, S.; Frati, G.; Vecchione, C.; Carrizzo, A. Targeting Nitric Oxide with Natural Derived Compounds as a Therapeutic Strategy in Vascular Diseases. *Oxidative Med. Cell. Longev.* **2016**, *2016*, 1–20. [CrossRef] [PubMed]

13. Akinmoladun, A.; Olowe, J.A.; Komolafe, K.; Ogundele, J.; Olaleye, M.T. Antioxidant Activity and Protective Effects of Cocoa and Kola Nut Mistletoe (Globimetula Cupulata) against Ischemia/Reperfusion Injury in Langendorff-Perfused Rat Hearts. *J. Food Drug Anal.* **2016**, *24*, 417–426. [CrossRef] [PubMed]

14. Komolafe, K.; Olaleye, T.M.; Omotuyi, O.I.; Boligon, A.; Athayde, M.L.; Akindahunsi, A.A.; Rocha, J.B. In Vitro Antioxidant Activity and Effect of Parkia biglobosa Bark Extract on Mitochondrial Redox Status. *J. Acupunct. Meridian Stud.* **2014**, *7*, 202–210. [CrossRef]

15. Caliceti, C.; Rizzo, P.; Cicero, A.F.G. Potential Benefits of Berberine in the Management of Perimenopausal Syndrome. *Oxidative Med. Cell. Longev.* **2015**, *2015*, 1–9. [CrossRef]

16. Tomas-Barberan, F.A. A New Process to develop a Cocoa Powder with Higher Flavonoid Monomer Content and Enhanced Bioavailability in Healthy Humans. *J. Agric. Food Chem.* **2007**, *55*, 3926–3935. [CrossRef]

17. Testai, L.; Martelli, A.; Cristofaro, M.; Breschi, M.C.; Calderone, V. Cardioprotective Effects of Different Flavonoids against Myocardial Ischaemia/Reperfusion Injury in Langendorff-Perfused Rat Hearts. *J. Pharm. Pharmacol.* **2013**, *65*, 750–756. [CrossRef]

18. Sanbongi, C.; Suzuki, N.; Sakane, T. Polyphenols in Chocolate, Which Have Antioxidant Activity, Modulate Immune Functions in Humans in Vitro. *Cell. Immunol.* **1997**, *177*, 129–136. [CrossRef]

19. Calderón, A.I.; Wright, B.J.; Hurst, W.; Van Breemen, R.B. Screening Antioxidants Using LC-MS: Case Study with Cocoa. *J. Agric. Food Chem.* **2009**, *57*, 5693–5699. [CrossRef]

20. Polson, C.; Sarkar, P.; Incledon, B.; Raguvaran, V.; Grant, R. Optimization of Protein Precipitation Based upon Effectiveness of Protein Removal and Ionization Effect in Liquid Chromatography-Tandem Mass Spectrometry. *J. Chromatogr. B* **2003**, *785*, 263–275. [CrossRef]

21. Di Paola, R.; Cordaro, M.; Crupi, R.; Siracusa, R.; Campolo, M.; Bruschetta, G.; Fusco, R.; Pugliatti, P.; Esposito, E.; Cuzzocrea, S. Protective Effects of Ultramicronized Palmitoylethanolamide (PEA-um) in Myocardial Ischaemia and Reperfusion Injury in Vivo. *Shock.* **2016**, *46*, 202–213. [CrossRef] [PubMed]

22. Ahmed, N.; Linardi, D.; Decimo, I.; Mehboob, R.; Gebrie, M.A.; Innamorati, G.; Luciani, G.B.; Faggian, G.; Rungatscher, A. Characterization and Expression of Sphingosine 1-Phosphate Receptors in Human and Rat Heart. *Front. Pharmacol.* **2017**, *8*, 312. [CrossRef] [PubMed]

23. Ben Mansour, R.; Gargouri, B.; Bouaziz, M.; Elloumi, N.; Jilani, I.B.; Ghrabi, Z.; Lassoued, S. Antioxidant Activity of Ethanolic Extract of Inflorescence of Ormenis Africana in Vitro and in Cell Cultures. *Lipids Heal. Dis.* **2011**, *10*, 78. [CrossRef] [PubMed]

24. Rizzo, A.; Mutinati, M.; Spedicato, M.; Minoia, G.; Trisolini, C.; Jirillo, F.; Sciorsci, R.L. First Demonstration of an Increased Serum Level of Reactive Oxygen Species During the Peripartal Period in the Ewes. *Immunopharmacol. Immunotoxicol.* **2008**, *30*, 741–746. [CrossRef]

25. Makara, M.A.; Hoang, K.V.; Ganesan, L.P.; Crouser, E.D.; Gunn, J.S.; Turner, J.; Schlesinger, L.S.; Mohler, P.J.; Rajaram, M. Cardiac Electrical and Structural Changes During Bacterial Infection: An Instructive Model to Study Cardiac Dysfunction in Sepsis. *J. Am. Hear. Assoc.* **2016**, *5*, 003820. [CrossRef]

26. Goya, L.; Martín, M.; Ángeles;Sarriá, B.; Ramos, S.; Mateos, R.; Bravo, L. Effect of Cocoa and Its Flavonoids on Biomarkers of Inflammation: Studies of Cell Culture, Animals and Humans. *Nutrients* **2016**, *8*, 212. [CrossRef]

27. Luo, Y.; Shang, P.; Li, D. Luteolin: A Flavonoid that Has Multiple Cardio-Protective Effects and Its Molecular Mechanisms. *Front. Pharmacol.* **2017**, *8*, 692. [CrossRef]

28. Serafini, M.; Bugianesi, R.; Maiani, G.; Valtuena, S.; De Santis, S.; Crozier, A. Plasma Antioxidants from Chocolate. *Nature* **2003**, *424*, 1013. [CrossRef]

29. Frangogiannis, N.G.; Dewald, O.; Xia, Y.; Ren, G.; Haudek, S.; Leucker, T.; Kraemer, D.; Taffet, G.; Rollins, B.J.; Entman, M.L. Critical Role of Monocyte Chemoattractant Protein-1/CC Chemokine Ligand 2 in the Pathogenesis of Ischemic Cardiomyopathy. *Circulation* **2007**, *115*, 584–592. [CrossRef]

30. Ahmed, N.; Linardi, D.; Muhammad, N.; Chiamulera, C.; Fumagalli, G.; Biagio, L.S.; Gebrie, M.A.; Aslam, M.; Luciani, G.B.; Faggian, G.; et al. Sphingosine 1-Phosphate Receptor Modulator Fingolimod (FTY720) Attenuates Myocardial Fibrosis in Post-Heterotopic Heart Transplantation. *Front. Pharmacol.* **2017**, *8*, 645. [CrossRef]

31. Sawyer, D.B.; Siwik, D.A.; Xiao, L.; Pimentel, D.R.; Singh, K.; Colucci, W.S. Role of Oxidative Stress in Myocardial Hypertrophy and Failure. *J. Mol. Cell. Cardiol.* **2002**, *34*, 379–388. [CrossRef] [PubMed]

32. Heo, H.J.; Lee, C.Y. Epicatechin and Catechin in Cocoa Inhibit Amyloid Beta Protein Induced Apoptosis. *J. Agric. Food Chem.* **2005**, *2005. 53*, 1445–1448. [CrossRef]

33. Doenst, T.; Nguyen, T.D.; Abel, E.D. Cardiac Metabolism in Heart Failure: Implications beyond ATP Production. *Circ. Res.* **2013**, *113*, 709–724. [CrossRef] [PubMed]

34. Abel, E.D.; Doenst, T. Mitochondrial Adaptations to Physiological vs. Pathological Cardiac Hypertrophy. *Cardiovasc. Res.* **2011**, *90*, 234–242. [CrossRef]

35. Santos, P.P.; Oliveira, F.; Ferreira, V.C.M.P.; Polegato, B.F.; Roscani, M.G.; Fernandes, A.A.; Modesto, P.; Rafacho, B.P.M.; Zanati, S.G.; Di Lorenzo, A.; et al. The Role of Lipotoxicity in Smoke Cardiomyopathy. *PLoS ONE* **2014**, *9*, e113739. [CrossRef] [PubMed]

36. Münzel, T.; Gori, T.; Keaney, J.; Maack, C.; Daiber, A. Pathophysiological Role of Oxidative Stress in Systolic and Diastolic Heart Failure and Its Therapeutic Implications. *Eur. Hear. J.* **2015**, *36*, 2555–2564. [CrossRef] [PubMed]

37. De Prati, A.C.; Podesser, B.; Faggian, G.; Scarabelli, T.; Mazzucco, A.; Darra, E.; Rungatscher, A.; Hallström, S.; Suzuki, H. Dual Modulation of Nitric Oxide Production in the Heart during Ischaemia/Reperfusion Injury and Inflammation. *Thromb. Haemost.* **2010**, *104*, 200–206. [CrossRef]

38. Rungatscher, A.; Hallström, S.; Linardi, D.; Milani, E.; Gasser, H.; Podesser, B.; Scarabelli, T.M.; Luciani, G.B.; Faggian, G. S-nitroso Human Serum Albumin Attenuates Pulmonary Hypertension, Improves Right Ventricular-Arterial Coupling, and Reduces Oxidative Stress in a Chronic Right Ventricle Volume Overload Model. *J. Hear. Lung Transplant.* **2015**, *34*, 479–488. [CrossRef]

39. Lustosa, B.B.; Polegato, B.F.; Minicucci, M.; Rafacho, B.; Santos, P.P.; Fernandes, A.A.; Okoshi, K.; Batista, D.; Modesto, P.; Gonçalves, A.; et al. Green Tea (Cammellia Sinensis) Attenuates Ventricular Remodeling after Experimental Myocardial Infarction. *Int. J. Cardiol.* **2016**, *225*, 147–153. [CrossRef]

40. Jose Corbalan, J.; Vatner, D.E.; Vatner, S.F. Myocardial Apoptosis in Heart Disease: Does the Emperor Have Clothes? *Basic Res. Cardiol.* **2016**, *111*, 31. [CrossRef]

41. Zhang, H.; Xue, G.; Zhang, W.; Wang, L.; Li, H.; Zhang, L.; Lu, F.; Bai, S.; Lin, Y.; Lou, Y.; et al. Akt and Erk1/2 Activate the Ornithine Decarboxylase/Polyamine System in Cardioprotective Ischemic Preconditioning in Rats: The Role of Mitochondrial Permeability Transition Pores. *Mol. Cell. Biochem.* **2014**, *390*, 133–142. [CrossRef] [PubMed]

MDPI

St. Alban-Anlage 66

4052 Basel

Switzerland

Tel. +41 61 683 77 34

Fax +41 61 302 89 18

www.mdpi.com

Antioxidants Editorial Office

E-mail: antioxidants@mdpi.com

www.mdpi.com/journal/antioxidants